CAPE CANAVERAL, FLORIDA

3…2…1… Lift-off! The state of Florida is home for much of NASA's space program. The Space Shuttle always lifts off from Cape Canaveral at NASA's Kennedy Space Center in Florida to start its amazing journey of discovery into outer space. The Space Shuttle is driven to the launch site by a special vehicle, a massive building on wheels.

One of Space Shuttle Endeavor's missions was to help repair NASA's Hubble Telescope. The Hubble Telescope has helped scientists look deep into space.

NASA's Space Shuttle program began sending astronauts into outer space in 1982. The Space Shuttle was scheduled to be taken out of service by NASA in September 2010.

NATIONAL GEOGRAPHIC

SCIENCE

FLORIDA

School Publishing

PROGRAM AUTHORS

Randy Bell, Ph.D.

Malcolm B. Butler, Ph.D.

Kathy Cabe Trundle, Ph.D.

Judith S. Lederman, Ph.D.

David W. Moore, Ph.D.

Program Authors

RANDY BELL, PH.D.

Associate Professor of Science Education,
University of Virginia, Charlottesville, Virginia
SCIENCE

MALCOLM B. BUTLER, PH.D.

Associate Professor of Science Education,
University of South Florida, St. Petersburg, Florida
SCIENCE

KATHY CABE TRUNDLE, PH.D.

Associate Professor of Early Childhood Science Education, The School of Teaching and Learning,
The Ohio State University, Columbus, Ohio
SCIENCE

JUDITH SWEENEY LEDERMAN, PH.D.

Director of Teacher Education, Associate Professor of Science Education, Department of Mathematics and Science Education,
Illinois Institute of Technology, Chicago, Illinois
SCIENCE

DAVID W. MOORE, PH.D.

Professor of Education,
College of Teacher Education and Leadership,
Arizona State University, Tempe, Arizona
LITERACY

Program Reviewers

Miranda Carpenter
Teacher, MS Academy Leader
Imagine School
Bradenton, FL

Kelly Culbert
K–5 Science Lab Teacher
Princeton Elementary
Orange County, FL

Richard Ellenburg
Science Lab Teacher
Camelot Elementary
Orlando, FL

Beth Faulkner
Brevard Public Schools
Elementary Training Cadre,
Science Point of Contact,
Teacher, NBCT
Apollo Elementary
Titusville, FL

Kathleen Jordan
Teacher
Wolf Lake Elementary
Orlando, FL

Melissa Mishovsky
Science Lab Teacher
Palmetto Elementary
Orlando, FL

Shelley Reinacher
Science Coach
Auburndale Central Elementary
Auburndale, FL

Flavia Reyes
Teacher
Oak Hammock Elementary
Port St. Lucie, FL

Rose Sedely
Science Teacher
Eustis Heights Elementary
Eustis, FL

Michelle Thrift
Science Instructor
Durrance Elementary
Orlando, FL

Cathy Trent
Teacher
Ft. Myers Beach Elementary
Ft. Myers Beach, FL

Acknowledgments
Grateful acknowledgment is given to the authors, artists, photographers, museums, publishers, and agents for permission to reprint copyrighted material. Every effort has been made to secure the appropriate permission. If any omissions have been made or if corrections are required, please contact the Publisher.

Illustrator Credits
All illustrations by Precision Graphics. All maps by Mapping Specialists.

Photographic Credits
Front Cover (fg) pbnj productions/ Photodisc/Getty Images. (bg) NASA/ Dimitri Gerondidakis.

Credits continue on page EM14.

National Geographic School Publishing
Hampton-Brown
www.myNGconnect.com

Printed in the USA.
RR Donnelley
Jefferson City, MO

ISBN: 978-0-7362-7816-4

10 11 12 13 14 15 16 17 18 19 20

1 2 3 4 5 6 7 8 9 10

LIFE SCIENCE

CONTENTS

TECHTREK
myNGconnect.com

Student eEdition | Vocabulary Games | Digital Library | Enrichment Activities

CHAPTER 2

How Do Parts of Living Things Work Together? 53

EARTH SCIENCE

CONTENTS

TECHTREK
myNGconnect.com

Student eEdition | Vocabulary Games | Digital Library | Enrichment Activities

CHAPTER 4

PHYSICAL SCIENCE

CONTENTS

CHAPTER 6

How Do You Describe Force and The Laws of Motion? 245

CHAPTER 7

How Do You Describe Different Forms of Energy? 277

TECHTREK
myNGconnect.com

Student eEdition

Vocabulary Games

Digital Library

Enrichment Activities

CHAPTER 8

FLORIDA

LIFE SCIENCE

What Is Life Science?

Life science is the study of all the living things around you and how they interact with one another and with the environment. This type of science investigates how living things are similar to and different from one another, how they live and reproduce, and how they function in the environment. Life science includes the study of humans, as well as all the other kinds of living things on Earth. People who study living things and the environment are called life scientists.

You will learn about these aspects of life science in this unit:

HOW DO LIVING THINGS SURVIVE AND CHANGE?

Plants and animals show a great variety of adaptations that enable them to survive in their environments. If the environment changes, those living things that cannot survive move to new environments or they die out. Life scientists study the adaptations of living things to understand how these organisms interact with the environment and with each other.

HOW DO PARTS OF LIVING THINGS WORK TOGETHER?

Humans have organs that work together to carry out life processes. Other kinds of living things also have organs that carry out similar tasks. Life scientists study the way the human body works and make comparisons to other living things.

NATIONAL GEOGRAPHIC

MEET A SCIENTIST

Maria Fadiman: Ethnobotanist

Maria Fadiman is an ethnobotanist and National Geographic Emerging Explorer. She was born with a passion for conservation and a fascination with indigenous cultures. Ethnobotany lets her bring it all together. Ethnobotany is the scientific study of the relationships that exist between people and plants.

Of her work, Maria says, "I used to think that going to the jungle made my life an adventure. However, after years of unusual work in exotic places, I realize that it is not how far off I go, or how deep into the forest I walk that gives my life meaning. I see that living life fully is what makes life—anyone's life, no matter where they go or do not go—an adventure."

On her first trip to the rain forest, Maria met a woman who was in terrible pain because no one in her village could remember which plant would cure her. Through this situation, Maria saw that knowledge of plant uses was being lost. In that moment she knew that learning more about indigenous people's use of plants was what she wanted to do with the rest of her life.

In Chapter 1, you will learn:

FLORIDA NEXT GENERATION SUNSHINE STATE STANDARDS

SC.5.L.15.1 Describe how, when the environment changes, differences between individuals allow some plants and animals to survive and reproduce while others die or move to new locations. **WHEN ENVIRONMENTS CHANGE**

SC.5.L.17.1 Compare and contrast adaptations displayed by animals and plants that enable them to survive in different environments, such as life cycles variations, animal behaviors, and physical characteristics. **PHYSICAL CHARACTERISTICS OF LIVING THINGS, BEHAVIORS HELP ANIMALS SURVIVE, LIFE CYCLE ADAPTATIONS**

SC.5.L.17.1 Science in a Snap! Compare and contrast adaptations displayed by animals and plants that enable them to survive in different environments, such as life cycles variations, animal behaviors, and physical characteristics.

CHAPTER 1

HOW DO LIVING THINGS SURVIVE AND

The red-eyed tree frog is well adapted to living in a tropical rain forest. Its long legs are good for climbing trees. On the end of each toe is a sucker pad that lets the frog stick to leaves and branches. Its green back makes it hard to see among the leaves. But the frog's bold red eyes can startle a predator and give the frog a chance to flee.

Red-eyed tree frogs live in the rain forests of Central America.

TECHTREK
myNGconnect.com

Student eEdition

Vocabulary Games

Digital Library

Enrichment Activities

CHANGE?

SCIENCE VOCABULARY

behavior (bi-HĀV-yur)

Behavior is any way that an animal interacts with its environment. (p. 16)

Hunting is a behavior that helps dragonflies survive.

instinct (IN-stinkt)

An **instinct** is an inherited behavior that an animal can do without ever learning how to do it. (p. 16)

Wasps build their nests by instinct.

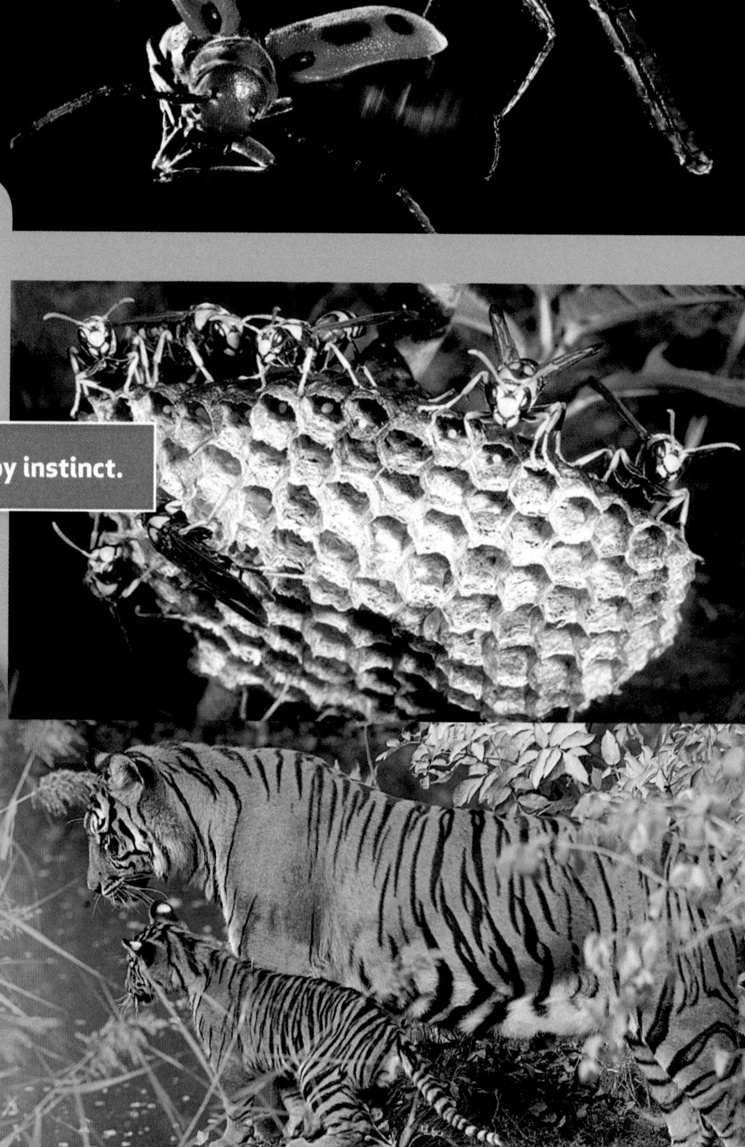

learning (LUR-ning)

Learning is a change in behavior that comes about through experience. (p. 17)

This tiger cub is learning how to fish by watching its mother.

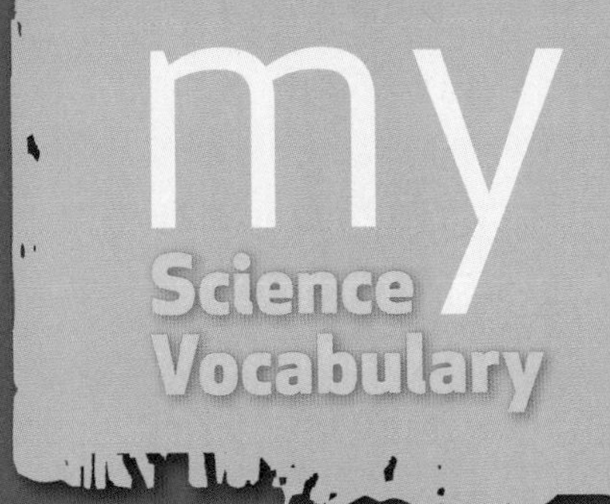

behavior
bi-HĀV-yur

communication
(kuh-MYŪ-ni-KĀ-shun)

habit
(HAB-it)

instinct
(IN-stinkt)

learning
(LUR-ning)

habit (HAB-it)

A **habit** is a behavior that is learned through practice. (p. 19)

This buffalo has developed the habit of not running away from snowmobiles.

communication (kuh-MYŪ-ni-KĀ-shun)

Communication is any behavior that lets animals share information. (p. 22)

Bird songs are one kind of communication in the animal world.

Physical Characteristics of Living Things

Animal Adaptations The many sharp teeth of the killer shark are there for a reason. The shark's name says it all. The shark's teeth make it a very good predator. They help the shark catch fish to eat.

The shark's teeth are an adaptation. Adaptations are characteristics that help organisms survive or reproduce. They are inherited characteristics that are passed from parents to offspring.

Like the killer shark, many other predators have sharp teeth to help them kill and eat other animals. In contrast, animals that eat plants have different kinds of teeth. For example, cows have teeth that are broad and flat. They use their teeth to grind up tough grasses, stems, and leaves. Many animals use their teeth for self-defense as well as for feeding.

This killer shark is well adapted for hunting prey.

The feet of animals are also adapted to different ways of life. Many animals have feet that help them find food or escape from predators.

Look at the animal feet in the chart. How do the feet of each animal help it survive?

COMPARING **ANIMAL FEET**

SHARP TALONS

Hawks have long sharp claws called talons. Hawks use their talons to capture and hold their prey.

SUCTION CUPS

The feet of lizards called geckos have tiny hairs on the bottom. These help their feet grip surfaces, allowing geckos to run up walls and across ceilings. This helps them capture insects and escape from predators.

STURDY HOOVES

The hooves of reindeer are hard and tough. Their hooves help them to run quickly over rocks and ice without getting hurt.

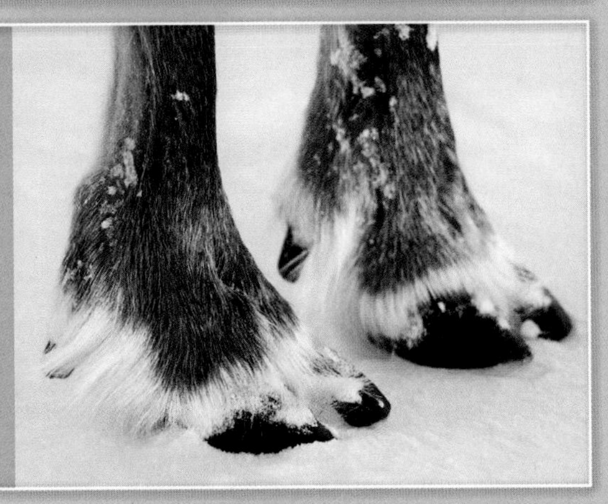

Animal Senses Most animals have senses such as sight, smell, and hearing. These senses are very important adaptations.

Some of the animals with the best sense of sight are predators that use their eyes to spot prey. Hawks are a good example. A hawk can see a rabbit up to a 1.5 kilometers (1 mile) away. That's why "hawk-eyed" means having keen eyesight.

Smell is another important sense in many animals. Moles and many other animals that live underground have very good senses of smell. They rely on smell to find food in the soil. Bears also have a good sense of smell. They depend on smell to find underground roots, insects, and bulbs. They also depend on smell to locate newly killed prey of other predators.

This mole uses its sense of smell to find food.

The red-tailed hawk has very good sight that it uses for finding prey.

A keen sense of hearing helps many animals survive. Some predators, such as owls, find food by listening for their prey. A good sense of hearing can also help an animal know when a predator is near. Deer use their large, sensitive ears to listen for predators. They can rotate their ears to pick up sounds that are coming from different directions.

A deer can move its ears in different directions. How does this adaptation help a deer survive?

Science in a Snap! Model Deer Ears

Work with a partner. Stand four meters apart. Listen as your partner says something in a soft voice. Then cup your hands behind your ears, and listen again.

Cup your hands in front of your ears, and listen to your partner again.

Which way was it easier to hear your partner? How is this activity like a deer listening for predators?

Plant Adaptations Plants have many adaptations that help them survive in their environments. Pitcher plants have an unusual adaptation. They are able to trap insects. They then digest the insects to get nutrients. This allows the pitcher plants to grow in places where the soil has few nutrients.

How do pitcher plants trap their prey? These plants have large, pitcher-shaped leaves with sticky liquid at the bottom. Insects slide down the slippery inside walls of the pitcher and drown in the liquid. The liquid slowly breaks down the dead insects and releases nutrients that the plant needs to survive.

This "pitcher" is an adaptation that allows the plant to capture insects.

Flowers are also examples of plant adaptations. Flowers attract pollinators such as insects. One way flowers attract pollinators is with bright colors. Different colors attract different animals.

Pollinators visit the flowers and carry pollen from one plant to another. Flowering plants must be pollinated in order to make seeds. Without pollinators, many plants could not reproduce.

Digital Library

FLOWER COLOR AND POLLINATORS

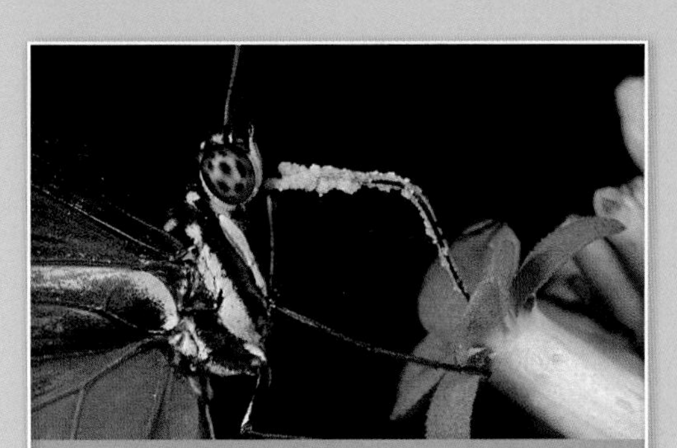

Orange flowers attract butterflies. This butterfly's long mouth part is covered with pollen.

Red flowers attract hummingbirds. Some also attract butterflies.

Yellow flowers attract bees. They may also attract other insect pollinators.

At night, white flowers are the most easily seen. They attract bats and moths that are active then.

Before You Move On

1. What are adaptations?
2. Explain how a bear's sense of smell helps it survive.
3. **Predict** A newly discovered plant has bright yellow flowers. What type of pollinator do you predict it will have? Why?

Behaviors Help Animals Survive

Instinct and Learned Behavior

The dragonfly on this page has a nickname. It is called the "pondhawk." The name fits because the pondhawk is a fierce predator. It chases and catches insects. It even preys on insects its own size.

Hunting is just one of many ways that animals behave in order to survive. **Behavior** is any way that an animal interacts with its environment. Besides finding food, behavior includes ways that animals protect themselves, make homes, find mates, and raise their young.

Animal behaviors may be either instinctive or learned. An **instinct** is a behavior that an animal can do without ever learning how to do it. Instincts are inherited characteristics. An instinct is always performed the same way. For example, dragonflies hunt by instinct. They always swoop down and capture their prey in the air. Most of the behaviors of insects are instinctive.

The green clearwing dragonfly on the right is chasing its prey—the red milkweed beetle on the left.

This Sumatran tiger is teaching her cub how to fish.

Behaviors that are not instincts must be learned. **Learning** is a change in behavior that comes about through experience. For example, a dog may learn to go to the back door when it wants to go outside. Learning is more important in animals with bigger brains. More intelligent animals can learn more, and more of their behavior is learned.

Young animals learn behaviors mainly from their parents. Look at the tigers in the picture. The cub is not just tagging along with its mother. It is learning how to find and catch fish by watching its mother do it.

Both instinctive and learned behaviors are important. Is one type of behavior better than the other? It depends on the situation.

Instinctive behaviors do not need to be learned, and they are always the same. However, if the environment changes and an instinctive behavior no longer works, the behavior cannot be changed.

It takes time and effort for an animal to learn a behavior. But learned behaviors are flexible. If the environment changes, a new behavior can be learned.

Protection Many behaviors help protect animals from predators. For example, opossums avoid predators by "playing dead." A frightened opossum goes limp and does not move. Predators usually leave it alone because they think it is dead. This is where the expression "playing possum" comes from.

The moth below has a behavior that helps scare away predators. When a predator approaches, the moth opens its wings. The two large spots on its lower wings look like the eyes of a fierce predatory bird. Spreading its wings for this purpose is an instinctive behavior.

Some animals have behaviors that help them hide from predators. Squid release a dark inky substance into the water when a predator is near. The ink blocks the predator's view of the squid, so the squid can quickly swim away.

When the moth's wings are closed, you cannot see its large spots. But when its wings are open, the spots look like an owl's eyes.

Many animals protect themselves by running away. A chipmunk darts under a rock to escape from a fox. A deer bolts through a field to escape from a chasing dog.

Other animals stand their ground and defend themselves. They may have characteristics such as sharp teeth, claws, or antlers that they can use to fight. They may also have characteristics that make them look large and fierce. Have you ever seen a cat raise its fur and arch its back when it is afraid? Reacting this way makes the cat look bigger and more dangerous.

Animals sometimes learn to stop running away from an "empty" threat. Look at the buffalo in the picture. It has learned that the snowmobiles are not dangerous, so there is no need to run away. This is an example of a **habit.** A habit is a behavior that is learned through practice. By staying put, the buffalo saves time and energy that can be put to better use, such as finding food.

A buffalo in Yellowstone National Park watches snowmobiles roar by.

Shelter and Raising Young Many behaviors provide shelter for animals or their young. Read about some of these behaviors in the chart.

Most animals do not take care of their young. But some animals, including birds and mammals, protect their offspring. Birds protect their eggs until they hatch. Then they feed their nestlings until they are big enough to get food for themselves.

Animals that live together in groups, such as paper wasps, often work together for the good of the group. Working together for the common good is called cooperation. Insects that live in colonies have many cooperative behaviors, such as building a nest and caring for young. These behaviors are instinctive.

Ring-tailed lemurs feed and protect their young. Lemurs live in Madagascar.

Migration Many kinds of animals migrate, or move to a different place when the seasons change. Migration is an instinctive behavior. In summer, many birds nest and raise their young in northern regions. Then they fly south for the winter. By migrating, these birds find enough food to survive.

Migrating animals don't have maps or GPS, so how do they know where to go? The position of the sun or landmarks such as rivers or mountains may help them find their way. Some animals can even sense the pull of Earth's magnetic North pole. It's as though they have a built-in compass to guide them.

ANIMAL **SHELTERS**

TECHTREK myNGconnect.com

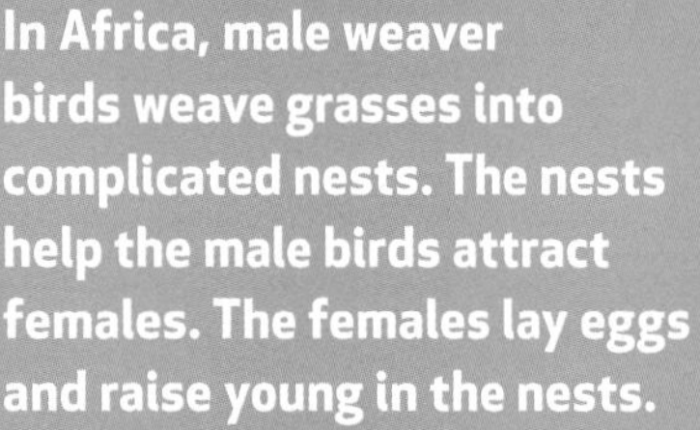

In Africa, male weaver birds weave grasses into complicated nests. The nests help the male birds attract females. The females lay eggs and raise young in the nests.

Pocket gophers dig burrows under the ground. They have separate rooms for sleeping and storing food. Each pocket gopher has its own burrow.

Paper wasps live in colonies. Members of a colony work together to build a nest. They use small pieces of plants and their own saliva to make the papery material.

Communication What do a hissing cat, a singing bird, and a dog that is wagging its tail have in common? All three are examples of animal **communication**. Communication is any behavior that lets animals share information.

Many animals communicate to attract mates and to protect themselves from predators. Animals may communicate with sounds, movements, smells, or by other means.

Look at the frilled lizard on this page. It communicates by flaring out a ruffle of skin around its neck. Can you guess what the lizard is communicating when it flares out its frill? It is saying, "I am a threat. Stay away!"

The frilled lizard is not as frightening when its frill is not flared.

Communication with sounds is important in the animal kingdom. Birds are well known for their songs. Male birds sing to attract mates. Singing is an instinctive behavior in birds. However, young birds may have to hear and practice their song before they get it exactly right.

Many birds use calls to warn each other of predators. Mammals such as meerkats and monkeys also use warning calls. They even have different warning calls for different predators, such as snakes or birds of prey. How might this information help members of the group survive?

This bokmakierie communicates by singing and making other sounds.

Before You Move On

1. What is an instinct?
2. How does learned behavior differ from instinctive behavior?
3. **Infer** How does communication help animals learn?

ALL HANDS ON DECK!
SAVING RIGHT WHALES

All of a sudden, a huge, dark creature shoots out of the ocean. Then it slaps the water with its flippers before disappearing again. It was a North Atlantic right whale! They aren't the largest type of whale. Even so, they are about the size of a school bus.

Right whales can dive deep into the ocean and hold their breath for up to 40 minutes. Like all whales, they are mammals, so they have to come back to the surface to breathe.

The name right whale comes from early whalers. They believed this was the "right" whale because it was easy to hunt. And hunt them they did! From the whales, they got oil for lamps and supports for umbrellas and fancy dresses. After hundreds of years of whaling, North Atlantic right whales were nearly extinct. By 1935 it was against the law to kill right whales. Fewer than 400 North Atlantic right whales are living today, making them the most endangered of all large whales.

Right whales are slow swimmers, but they are like acrobats when they jump out of the water.

North Atlantic right whales live along the East Coast of North America. During summer and fall, many gather to feed in the Gulf of Maine and the Bay of Fundy in Canada.

In early winter, female right whales migrate to the warmer waters off the coast of Georgia and Florida. There they give birth. Mothers stay close to their calves to nurse and protect them. During April, the mothers and calves swim north to join other right whales in feeding grounds in the Bay of Fundy and off the coast of Massachusetts.

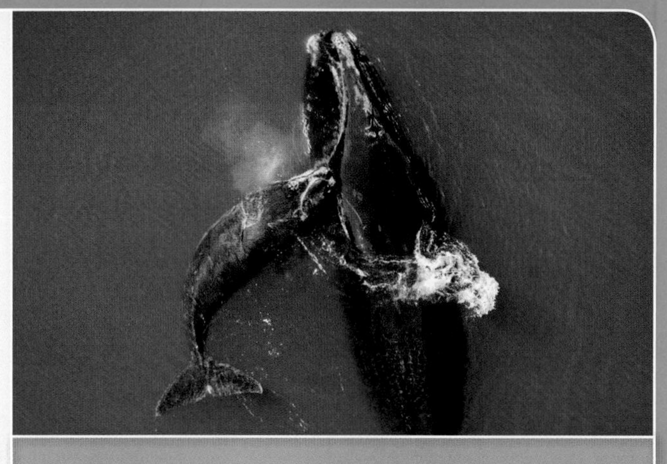

Female right whales give birth to a single calf.

Right whales use the baleen in their mouths to strain tiny animals from the ocean.

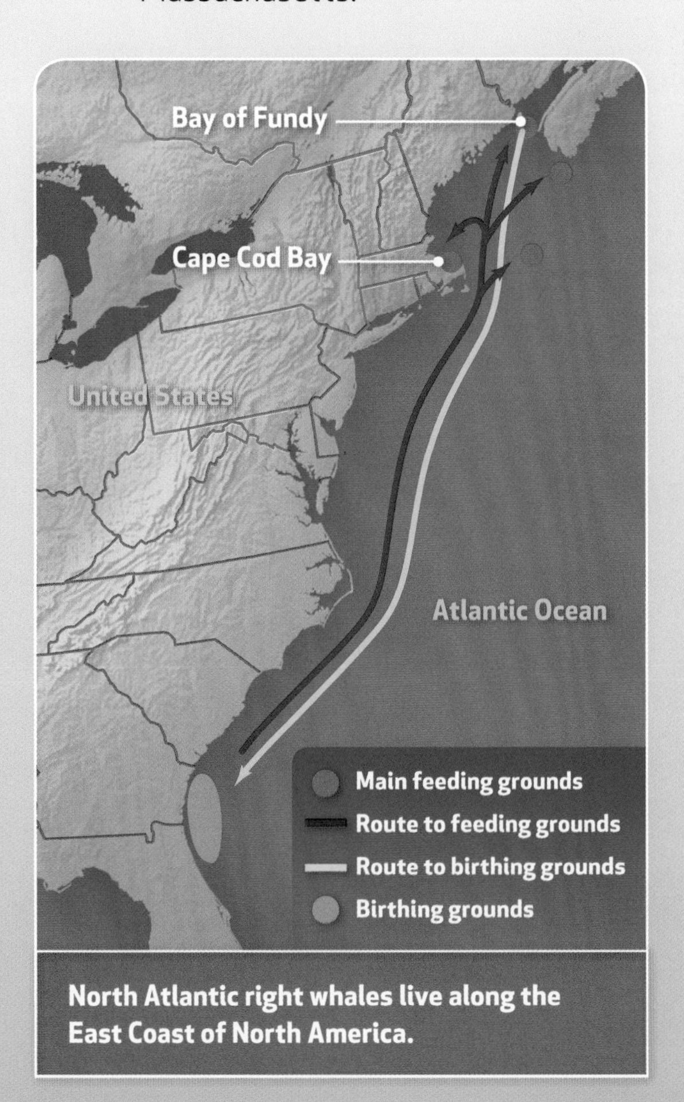

North Atlantic right whales live along the East Coast of North America.

Right whales have special adaptations for feeding. Instead of teeth, they have baleen. Baleen is made of the same material as your fingernails. Hundreds of rows of baleen hang from the whales' upper jaws. The whales feed by swimming slowly with their mouths open. As water passes through the rows of baleen, tiny ocean organisms are trapped. Right whales can eat more than a ton of these tiny creatures in a single day.

Even though right whales are now safe from whalers, they face other serious threats. As they migrate up and down the Atlantic coast, they stay fairly close to shore. Many ships also travel through these waters. Collisions with ships are the biggest killer of right whales. Whales that do not die are often injured and scarred.

Sometimes right whales are caught in fishing ropes and nets. As they try to get free, the ropes cut through their skin. Sometimes the tangled ropes keep them from feeding, and they eventually starve.

Ocean pollution and smaller food supplies also threaten right whales. Some scientists believe that without protection, right whales could become extinct in less than 200 years.

The good news is that many people are working together to protect the right whales. Scientists, government agencies, and the fishing industry are finding ways to save whales. Preventing the deaths of just two adult females each year would give the right whales a good chance to recover.

One of the greatest threats to right whales is the ropes, nets, and traps used by fishing boats.

For 30 years, a research team from Boston's New England Aquarium has been studying right whales. These scientists have gathered almost 400,000 photographs to help them identify nearly every North Atlantic right whale. Each whale is given a number, and sometimes a name. This information helps scientists track individual whales.

Tracking individual whales helps scientists learn more about the right whale population so they can plan for its recovery.

Every sighting of a right whale is sent to a warning system that lets ships know whales are nearby. U.S. and Canadian officials have lowered ship speeds and changed shipping routes where right whales are found.

New fishing rules are helping protect the whales. Fishing ships cannot use certain kinds of gear that are harmful to whales. Researchers are working on fishing lines that will break if whales get caught in them.

There are also rescue teams that work to free whales that are tangled up in fishing lines.

Are these measures helping the whales? Since 2001, the number of right whales has been increasing. In 2009, a record of 39 calves were born. And there have been far fewer deaths caused by ships and fishing gear. These are encouraging signs of recovery. But most scientists say the whales are not out of danger yet.

The white scars on its tail help scientists identify this right whale.

Life Cycle Adaptations

Plant Life Cycles The wildflowers on this page are growing in a desert. Does that surprise you? These plants have adaptations that help them survive in a dry environment. One adaptation is their ability to go through their entire life cycle very quickly after it rains. In just a few weeks, seeds germinate, stems and leaves grow, flowers bloom, and seeds form.

The plants spend the rest of the time as seeds. Their seeds can survive dry conditions for months, or even years, until it rains again.

In many climates, the weather is too cold for plants to grow during the winter. In most of North America, plants can grow only during spring, summer, and fall. Many flowering plants die when the weather turns cold. When spring returns, new plants must grow from seeds again. Plants with this type of life cycle may produce very large numbers of seeds each year.

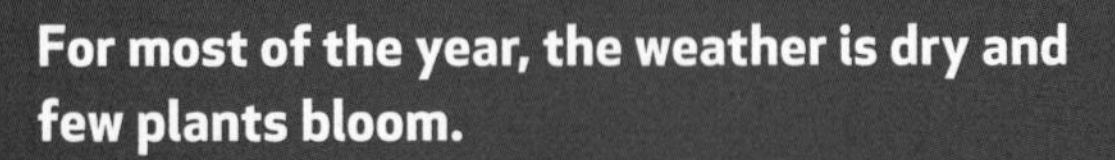

For most of the year, the weather is dry and few plants bloom.

When rain comes to Picacho Peak State Park in Arizona, the land is covered with wildflowers.

Other kinds of plants live for many years. These plants may go dormant each winter and lose their leaves or even die back to the ground. In spring, they start growing again. Some plants do not lose their leaves during the winter. These plants are called evergreens. Pine trees are an example.

Many plants survive winter or a dry season by storing food in underground bulbs. Plants that form bulbs include hyacinths and tulips. The pictures below show the hyacinth's yearly growing cycle.

YEARLY CYCLE OF A **HYACINTH**

A hyacinth is a plant that grows from a bulb and blooms in the spring.

Animal Life Cycles The life cycles of animals help them survive in their environment. Insects have some of the most interesting life cycles. You can see the life cycle of the monarch butterfly on the next page. Notice that a monarch has four stages: egg, larva, pupa, and adult.

In each stage of its life cycle, an insect has different needs. For example, larvae need a great deal of energy to grow and change into adults. They get the energy they need by eating constantly, day and night. The larvae hungrily munch their way through leaf after leaf. Insects must go through their larval stage at a time of year when there are plenty of leaves. In most of the United States, this is during the spring or summer.

Insects that live in places with cold winters must have adaptations that protect them from freezing. These adaptations vary, depending on the stage at which insects go through the cold season. Many insects spend the winter as eggs. The eggs do not hatch until the weather is warm enough for the larvae to live.

Insects that spend the winter as pupae are protected inside a chrysalis or cocoon. Some adult insects find a protected place and spend the winter in a sleep-like state, similar to the hibernation of other animals. Insects that live in colonies crowd together in their nest to share body heat.

A few adult insects migrate to a warmer climate. For example, monarch butterflies migrate from Canada and the northern United States to spend the winter in Mexico.

Enrichment Activities

LIFE CYCLE OF A **MONARCH**

Monarch butterflies go through a life cycle called complete metamorphosis.

LARVA Butterfly larvae are called caterpillars. Each caterpillar eats and grows larger.

PUPA The caterpillar forms a protective chrysalis. Inside the chrysalis, the pupa goes through many changes.

ADULT The adult has wings. It is ready to find a mate and reproduce.

EGGS Monarchs lay their eggs on milkweed plants.

Before You Move On

1. List three adaptations that protect adult insects from the cold.
2. Compare the life cycle of a plant that lives for just one year with the life cycle of a plant that lives for many years.
3. **Apply** Many mammals such as groundhogs hibernate, or sleep deeply during the winter. How does this behavior help the groundhogs survive?

When Environments Change

Differences Among Individuals These wild horses all belong to the same species. In many ways they are just alike. For example, they all have pointed ears, two eyes, and a mane. But the horses are not alike in every way. What differences do you see?

The most obvious difference is the color of their hair. But the horses also differ in many less obvious ways , such as their strength and speed.

The individual members of all species differ in many ways. Some of these differences affect their ability to survive. For example, wild horses that can run faster may be better able to escape from predators.

Although all the horses in this herd are similar, individual horses differ from each other.

Plants as well as animals show variation in their characteristics. Compare the two bush lilies shown here. Both plants belong to the same species, but their flowers are different colors. Flower color can affect a plant's ability to reproduce, because colorful flowers attract pollinators such as insects.

Yellow bush lilies and red bush lilies are found in different places. Yellow bush lilies are found where bees are common pollinators. What pollinators do you think are common where red bush lilies are found?

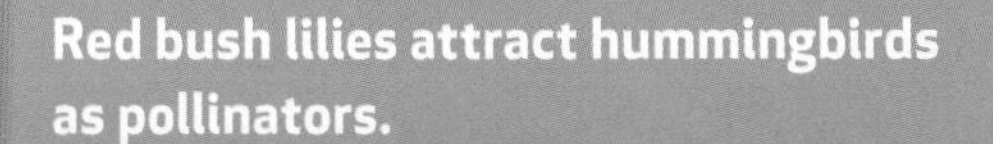

Red bush lilies attract hummingbirds as pollinators.

Yellow bush lilies attract bees as pollinators.

Changes in Populations

You can't see them in the big picture, but many animals live in this desert. Rock pocket mice live in small burrows near the rocks.

Most rock pocket mice have brown fur, but some have black fur. In most places, brown mice are more common than black mice. But in places with black rocks, black mice are more common. Can you explain why?

On black rocks, it is hard for predators to see black mice. Their black fur blends in with the rocks. Since they are harder to catch, black mice are more likely to survive than brown mice. Therefore black mice live longer and have more offspring. Over time, mice with black fur become more common.

When some individuals of a population are better adapted than others, their characteristics can be passed on to the next generation. Over time, their characteristics become more common in the population.

Brown rock pocket mice blend in with light-colored rocks. This helps them hide from predators.

On dark-colored rocks, it is harder to see black rock pocket mice.

When environments change, differences among individuals may allow some organisms to survive while others die. Scientists observed this happening on the Galápagos Islands, which are in the Pacific Ocean.

Some Galápagos finches eat seeds, which they crack open with their beaks. Birds with bigger beaks can crack open bigger seeds. During several dry years, seeds were scarce. Then many birds with smaller beaks could not find enough small seeds to survive.

Birds with bigger beaks could crack and eat more seeds. Most of these birds survived and reproduced. In just a few generations, most of the finches had larger beaks.

Some organisms survive change by moving to a new location. Recently, summers in the Arizona desert have become hotter and drier. Some plants cannot live where it is so hot and dry. But seedlings of these plants now grow higher up the mountainside. How does this help? Higher on the mountains the weather is cooler and wetter, so the plants can survive.

Finches with small beaks can crack open only small seeds.

Finches with large beaks can crack open both large and small seeds.

How Environments Change

Environments can change in many ways. For example, the habitat where an organism lives may get smaller, or a new organism may move into the habitat. Some living things have characteristics that let them survive such changes. Others do not.

The habitats of many species are getting smaller. For example, new buildings, roads, and farms have taken away habitat from the Florida panther. These cats need large areas to hunt and find prey.

Once, thousands of panthers roamed the entire state. Today, their only habitat is at the tip of South Florida. Fewer than 100 panthers are living in Florida now.

When a new disease comes into an area, it can cause big problems. In the 1900s, wood from Europe brought Dutch elm disease to North America. Within 50 years, more than 100 million American elm trees were killed by this disease. But a few American elm trees had characteristics that let them resist the disease. Those trees have been able to grow and reproduce. In the future, trees that can resist the disease may become more common.

The Florida panther has lost habitat, and is now found only in a small area of South Florida.

A new species can upset the balance of an ecosystem. One species of fish that is new to the Mississippi River is changing that river's ecosystem. Asian carp were brought from Asia to help keep fish ponds clean. But floods washed the carp into the Mississippi River. Asian carp reproduce quickly and have no predators here, so their population grew quickly. Soon they were crowding out native fish.

Pollution can harm the living things in an environment. You may think the ocean is so huge that pollution couldn't cause a problem. But harmful chemicals and trash have spread to all parts of the ocean. Many fish and other organisms are being poisoned by the polluted water they live in. Sea birds, such as pelicans, are also poisoned when they eat the fish.

Asian carp are big fish with big appetites, but they scare easily. A passing boat can cause them to leap out of the water like torpedoes, sometimes hurting people.

Sometimes ships carrying oil crash, spilling the oil into the ocean. The oil harms plants and animals and spoils beaches.

Extinction When environments change, organisms that cannot adapt may go extinct. This happened to the dusky seaside sparrows. These sparrows used to live in the salt marshes of Merritt Island in Florida. They began dying off in the 1940s, when people sprayed poison on marshes to kill mosquitoes. The birds were poisoned when they ate the insects. The poison was banned in the 1970s, but then people started draining the marshes. This destroyed most of the bird's habitat. The species was declared extinct in 1990.

Many species have recently gone extinct, and many more species are at risk of extinction in the future. Extinction is nothing new. As many as 99 percent of all species that ever lived on Earth have gone extinct. But recently the rate of extinction has been increasing. Pollution, the introduction of new species and diseases, and habitat loss are the main causes of modern extinctions.

The dusky seaside sparrow used to live on Merritt Island in southern Florida salt marshes before it went extinct.

Merritt Island National Wildlife Refuge, Florida

Animal extinctions get more attention, but many plants have also gone extinct. One example is the Franklin tree, which was named for Benjamin Franklin. It grows in gardens but no longer grows in the wild. The rise of cotton farming in Georgia, where the tree used to grow, may explain why. A fungus that grows on cotton plants also attacks and kills the tree.

Humans have protected other species from going extinct. The bald eagle almost died out in the 1960s. The same poison that killed the dusky seaside sparrow was killing eagles. After the poison was banned, the bald eagle started to make a comeback. By the 1990s, it was no longer considered in danger of extinction.

Sometimes organisms thought to be extinct have later been found alive. The coelacanth fish was thought to have gone extinct 65 million years ago, But in 1938, scientists were surprised when a living coelacanth was found. Since then, several more have been discovered.

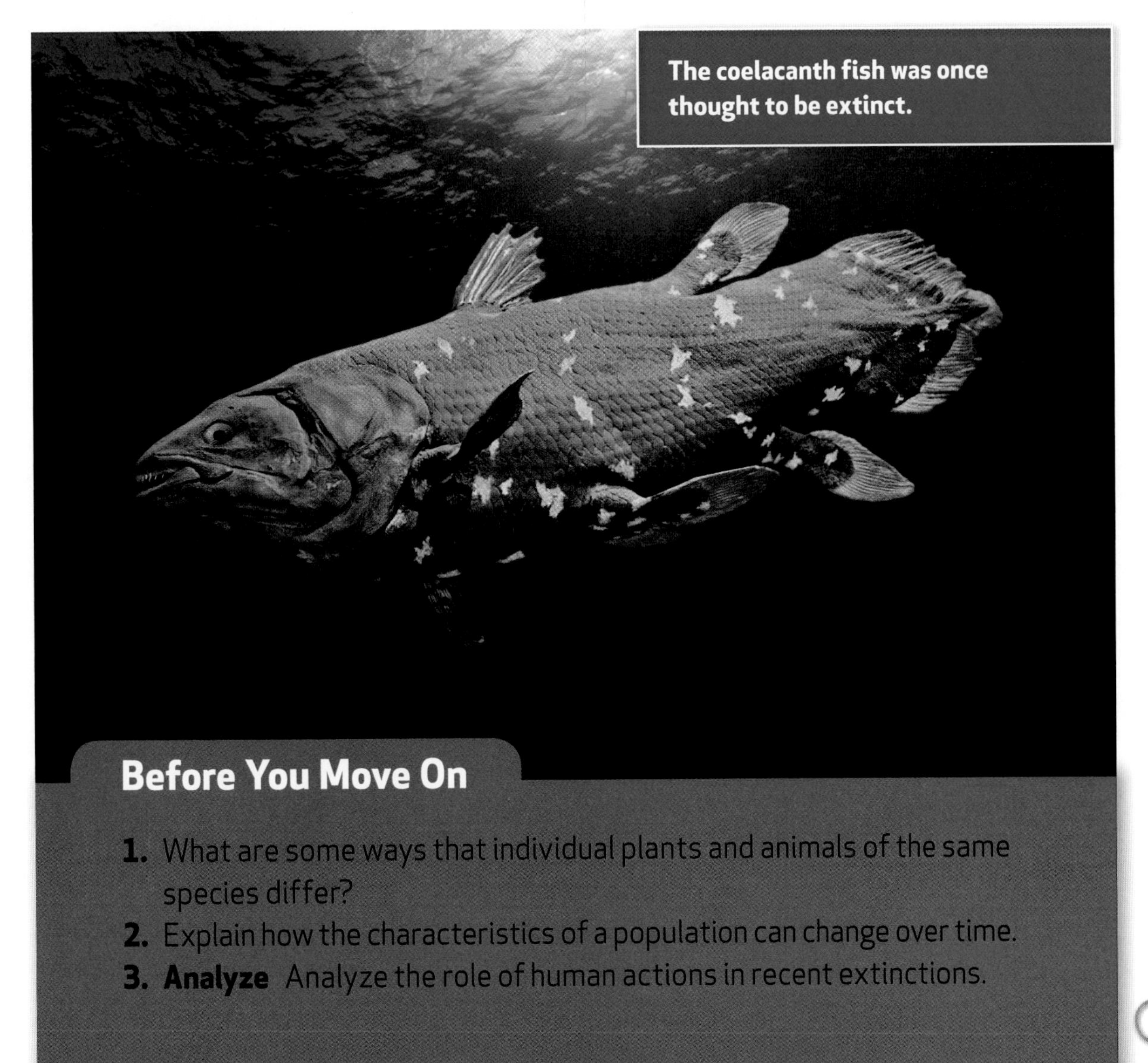

The coelacanth fish was once thought to be extinct.

Before You Move On

1. What are some ways that individual plants and animals of the same species differ?
2. Explain how the characteristics of a population can change over time.
3. **Analyze** Analyze the role of human actions in recent extinctions.

Conclusion

Living things have many different adaptations that help them survive. Most adaptations are inherited. Animals depend on behavior to survive. Their behavior may be instinctive or learned. Some animals have behaviors that let them communicate information to other animals. The life cycles of plants and animals are also adaptations that aid in survival. When environments change, differences among individual organisms allow some to survive while others die or move to other locations.

Big Idea Organisms have many different kinds of adaptations that let them survive and reproduce in their environment.

Adaptations include physical characteristics and behaviors.

Vocabulary Review

Match each of the following terms with the correct definition.

A. behavior
B. communication
C. habit
D. instinct
E. learning

1. Behavior that is learned through practice
2. Any behavior that lets animals share information
3. Change in behavior that comes about through experience
4. Any way that an animal interacts with its environment
5. Inherited behavior than an animal can do without ever learning how to do it

Big Idea Review

1. **List** What are three ways that animals use communication?
2. **Describe** Describe an adaptation that helps protect some insects from the cold.
3. **Relate** How do the colors of flowers relate to pollination?
4. **Cause and Effect** How does migration help animals survive?
5. **Infer** Assume that all the members of a species are just alike. How do you think this would affect their ability to survive if their environment changes?
6. **Make Judgments** What responsibility do you think human beings have to protect other living things from extinction?

Write About Animal Behavior

Explain Look at the weaver bird in the picture. Describe what he is doing. Explain how his behavior helps him reproduce.

CHAPTER 1 LIFE SCIENCE EXPERT: AQUATIC ECOLOGIST

Zeb Hogan

Dr. Zeb Hogan leads the National Geographic MegaFishes Project, which works to study and protect Earth's largest freshwater fishes and the rivers where they live. Zeb is an aquatic ecologist at the University of Nevada — Reno. He has studied fishes on six continents—North America, South America, Europe, Africa, Asia, and Australia.

What does an aquatic ecologist do?

Aquatic ecologists study the interplay between aquatic organisms, such as fish, and their environment. My specialty is megafish. Megafish are freshwater fish that grow more than 6 feet (1.8 m) long or weigh more than 200 pounds (91 kg). There are about two dozen species of megafish in the world.

What is a typical day for you in the field?

In the field, I spend most of my time on or near the water. In Southeast Asia, I rely on fishers to help me gather information. We ask them about the number and kinds of fish they are catching. If they catch an endangered fish, we help them tag it and release it back into the water.

I also work with National Geographic to produce news stories and shows for television. This gives me a chance to share my experiences with many people.

Zeb Hogan with a taimen in Mongolia

Zeb shows off a giant stingray in Cambodia.

Can you judge the size of this taimen by looking at Zeb's outstretched arms?

What has been your greatest accomplishment so far?

I'm most proud of my work with the Mekong giant catfish. I was part of a team that helped to list the catfish as Critically Endangered. Since that listing, we've convinced some fishers to stop fishing for the catfish. We are now working with other scientists and government officials to find the best ways to protect the species.

Why is your work important?

Some megafish, such as sturgeon, have lived on Earth for over 100 million years. I feel strongly that these fish and the rivers where they live need our help. The conservation of these unique species is important. They are indicators of the health of freshwater ecosystems.

NATIONAL GEOGRAPHIC

BECOME AN EXPERT

The Amazing World of Ants

Ants All Around Us Imagine trying to pick up and carry a car—with your jaws! Humans cannot lift things that are so much heavier than themselves, but ants can. Ants can pick up things that are many times their own weight and carry them long distances. Ants may be small, but they are mighty.

These ants are holding onto each other to form a living ladder.

There are many different kinds of ants—more than 10,000 species. Ants live just about everywhere on land except near the poles. In some places, one out of every two insects is an ant. Each ant is very small, but there are huge numbers of them. Together, they make up 20 percent of the total mass of land animals. In fact, ants have a greater mass than all land vertebrates combined!

The key to ants' success is their social behavior. All ants live in groups called colonies. Ant colonies range in size from dozens to millions of ants.

behavior

Behavior is any way that an animal interacts with its environment.

TECHTREK
myNGconnect.com

Student eEdition

Digital Library

Through instinct, members of a colony cooperate in everything they do. By working together, they can do many things that a single ant could never do alone. The ant ladder in the picture is a good example. Other ants can scramble up the ladder to reach a high branch that a lone ant would not be able to reach.

Most ants will eat just about anything, but some prefer certain foods, such as insects or seeds. Many ants like sweet liquids, such as plant sap or honeydew, a sweet liquid secreted by insects called aphids. Some ants even "herd" aphids to have a steady supply of honeydew. Only a few ant species, such as leafcutter ants, eat a single kind of food. Leafcutter ants feed only on fungus. They grow the fungus on beds of chewed leaves in their nest.

Leafcutter ants carry leaves back to their nest.

instinct

An **instinct** is an inherited behavior that an animal can do without ever learning how to do it.

Ant Castes

Like many other insects, ants undergo complete metamorphosis. They go through four life stages: egg → larva → pupa → adult.

Each adult belongs to one of three different classes, or castes: queen, drone, or worker. Each caste looks different from the others and has a different job in the colony.

The queen is the head of the colony. Her only job is to lay eggs. She may lay up to 1000 eggs a day, day after day for many years. She needs to mate just once to have all of her eggs fertilized. A queen has wings until she mates. After that, she breaks off her wings and stays in the nest.

Drones are the only males in the colony. Drones do not do any work. After they mate with a queen, they die.

TECHTREK
myNGconnect.com
Digital Library

A queen fire ant with pupae and workers

In many ant species, there are different types of workers. For example, some workers may have adaptations to be soldiers. The job of soldiers is to defend the nest from predators. What characteristics make the soldier ant pictured here well suited for this job?

This soldier ant has large, powerful jaws. It uses its jaws to bite predators.

In a few ant species, some workers are even more specialized. For example, in honeypot ants, some workers have adaptations for storing food inside their body (see picture).

An ant's caste depends mainly on the food it gets as a larva. Larvae that get the food with the most nutrients develop into queens. Others develop into drones or workers. Because caste is affected by the environment in this way, an ant's caste is not fixed when it hatches. This makes ant colonies flexible. They can adjust the numbers in each caste to meet changing needs. This improves the colony's chances of surviving.

These honeypot ants eat and store sweet liquids inside their body until they swell to the size of a marble. They spit up and share some of the liquid when other workers tap them with their antennae.

Ant Homes

Once a year, new adult drones and queens emerge from the pupa stage. They soon fly into the air and mate. Then the drones die, while the queens start new colonies. Each queen digs a small nest and lays her first batch of eggs. As the eggs hatch and complete the life cycle, worker ants emerge. The workers expand the nest and take over all the other work of the colony.

Most ants dig underground nests with many chambers, or rooms, connected by tunnels. Some chambers are used for nurseries, where eggs, larvae, or pupae are cared for. Other chambers are used for storing food or for egg laying by the queen. In cold climates, there may be chambers deep below the surface where ants can go in the winter to stay warm. Ant nests may be very deep. Some reach more than 10 meters (35 feet) below the surface.

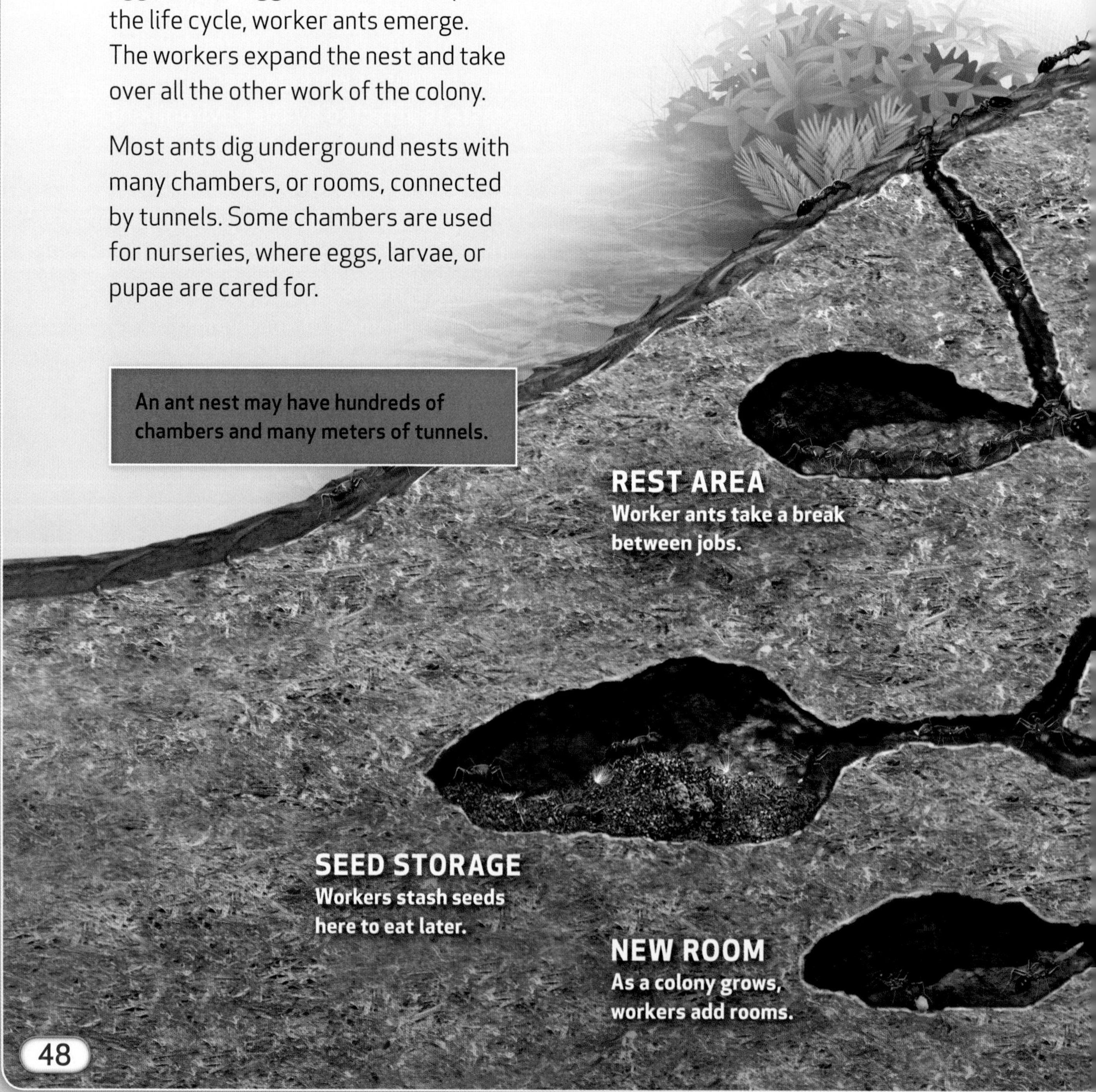

An ant nest may have hundreds of chambers and many meters of tunnels.

REST AREA
Worker ants take a break between jobs.

SEED STORAGE
Workers stash seeds here to eat later.

NEW ROOM
As a colony grows, workers add rooms.

Not all ants dig underground nests. Some make nests in trees or dead logs. A few species do not build nests at all. Army ants, for example, travel all day, eating just about everything in their path. Each night, they cluster together in a different tree.

An ant colony may last for many years and grow to have millions of members. Queens can live for up to 30 years. Most colonies last until the queen dies.

Without a queen, the colony will not last long. The worker ants die off, and there are no new workers to replace them.

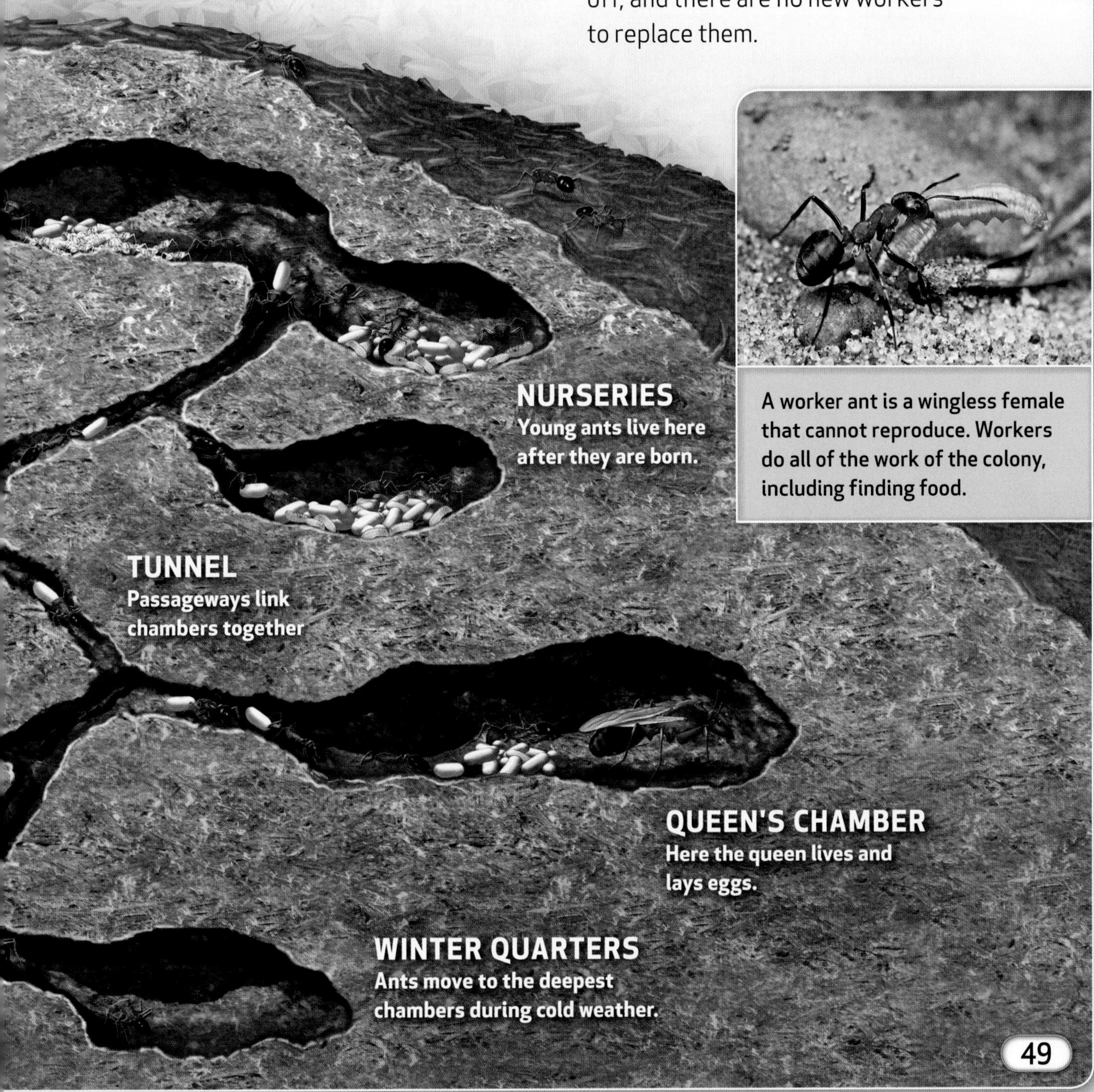

A worker ant is a wingless female that cannot reproduce. Workers do all of the work of the colony, including finding food.

NURSERIES
Young ants live here after they are born.

TUNNEL
Passageways link chambers together

QUEEN'S CHAMBER
Here the queen lives and lays eggs.

WINTER QUARTERS
Ants move to the deepest chambers during cold weather.

Ant Communication and Learning

Ants could not live and work together in such large numbers without forms of communication. The main way ants communicate is with chemicals. In fact, they have the most complex chemical communication in the animal kingdom. They give off chemicals, called pheromones, from special glands. They smell the pheromones of other ants with their antennae.

Different ant pheromones mean different things. For example, some pheromones mean "danger." Other pheromones show which colony an ant belongs to. Still others are used to mark trails. Ants may travel up to 200 meters (700 feet) from their nest looking for food, so marking trails is important. It helps them find their way back to the nest, and it shows other ants where to find food.

These ants are using their antennae to identify each other as nestmates.

communication

Communication is any behavior that lets animals share information.

Ants also communicate with sound, touch, and taste. They make sounds by rubbing their mouth parts on their body. They don't have ears, but they can sense sounds through their feet and antennae. Ants also use their antennae to feel things, especially in their dark underground nests. They use their antennae to taste things as well. No wonder ants always seem to be moving their antennae! It's not a **habit**. It's their instinctive way of sensing the world.

Most ant behaviors are instinctive, but **learning** also takes place in ants. For example, older workers have been seen teaching younger workers where to find food. The teachers lead the students along a trail to a food source. They slow down whenever the students lag behind, until the students catch up.

Unlike most other insects, ants are able to bend their antennae. This makes their antennae more useful for sensing the environment.

habit

A **habit** is a behavior that is learned through practice.

learning

Learning is a change in behavior that comes about through experience.

NATIONAL GEOGRAPHIC

CHAPTER 1

SHARE AND COMPARE

Turn and Talk How does living in colonies help ants survive? Form a complete answer to this question together with a partner.

Read Select two pages in this section. Practice reading the pages. Then read them aloud to a partner. Talk about why the pages are interesting.

Write Write a conclusion that tells the important ideas you learned about the behavior of ants. State what you think is the Big Idea of this section. Share what you wrote with a classmate. Compare your conclusions.

Draw Imagine what it would be like to be a worker ant, living in a colony with many other ants. Decide on the type of worker you would be and draw a picture of an activity you might do. Combine your drawing with those of your classmates to make an ant colony mural.

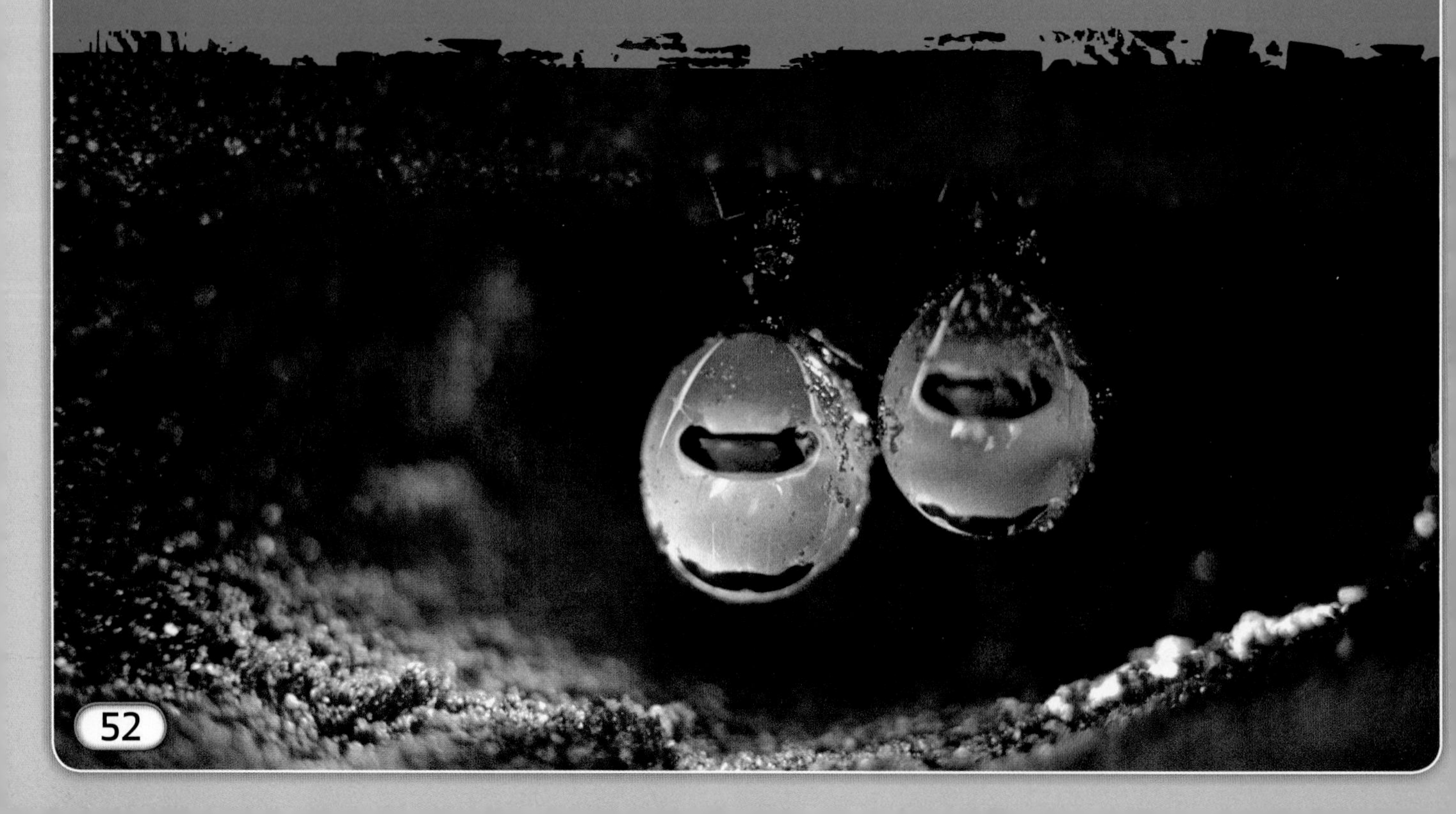

In Chapter 2, you will learn:

FLORIDA NEXT GENERATION SUNSHINE STATE STANDARDS

SC.5.L.14.1 Identify the organs in the human body and describe their functions, including the skin, brain, heart, lungs, stomach, liver, intestines, pancreas, muscles and skeleton, reproductive organs, kidneys, bladder, and sensory organs. **ORGANS IN THE HUMAN BODY, MOVING THE HUMAN BODY, MOVING BLOOD AND OXYGEN, GETTING FOOD AND REMOVING WASTES, CONTROLLING THE HUMAN BODY**

SC.5.L.14.2 Compare and contrast the function of organs and other physical structures of plants and animals, including humans, for example: some animals have skeletons for support (some with internal skeletons others with exoskeletons). **ORGANS IN PLANTS AND OTHER ANIMALS**

SC.5.L.14.1 Science in a Snap! Identify the organs in the human body and describe their functions, including the skin, brain, heart, lungs, stomach, liver, intestines, pancreas, muscles and skeleton, reproductive organs, kidneys, bladder, and sensory organs.

CHAPTER 2

HOW DO PARTS OF LIVING THINGS WORK

Even when you're asleep, your body is working. So think about how busy the different parts of this person's body are! What parts seem to be working the hardest? Compare her action to what you're doing right now. What parts of your body are working the hardest?

TECHTREK
myNGconnect.com

Student eEdition

Vocabulary Games

Digital Library

Enrichment Activities

THINGS TOGETHER?

The ballerina slips to the floor and holds her arms in this pose. How does her body do this?

SCIENCE VOCABULARY

organ (ŌR-gun)

An **organ** is a structure that carries out a specific job in the body. (p. 58)

The skin is an organ.

artery (AR-tur-ē)

An **artery** is a blood vessel that carries blood away from the heart. (p. 64)

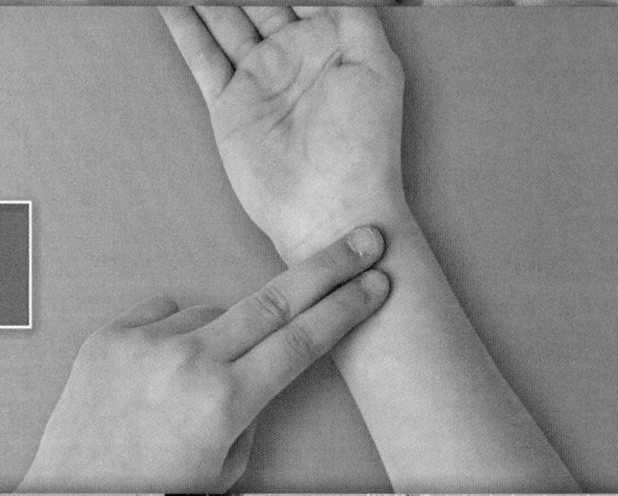

You feel an artery under the skin when you feel your pulse.

vein (VĀN)

A **vein** is a blood vessel that carries blood toward the heart. (p. 64)

Blood with waste products in it moves through the boy's veins.

artery (AR-tur-ē)

kidney (KID-nē)

liver (LIV-ur)

organ (ŌR-gun)

pancreas (PĀN-krē-us)

vein (VĀN)

Vocabulary Games

pancreas (PĀN-krē-us)

The **pancreas** is an organ that helps digest proteins and starches. (p. 71)

The pancreas adds juices to the small intestine to digest food.

liver (LIV-ur)

The **liver** is an organ that produces bile to help digest fatty foods. (p. 71)

liver

The liver works hard when you eat fatty foods.

kidney (KID-nē)

A **kidney** is an organ that removes wastes and extra water from the blood. (p. 73)

The kidneys are located in your lower back.

Organs in the Human Body

The human body is made up of dozens of **organs**. These structures carry out specific jobs as you live and grow. Your skin is your largest organ. It covers your entire body. Unlike most other organs that stay about the same size throughout your life, the body adds more skin as you grow larger.

Your skin has certain jobs. Besides protecting everything inside your body, it helps you cool down when you get hot. Also, skin helps your body get rid of wastes that are produced when you perspire. The skin includes parts that grow hair and that enable you to feel hot and cold.

These runners feel the cool breeze because of structures in their skin!

You can't see most of your organs. They're tucked away inside your body. Your brain is covered by the bones of your skull. Other organs, such as your stomach, are protected by thick layers of muscles. The photo to the right was made by a special machine that uses magnets and radio waves to make images. Images such as these help determine whether an organ is working correctly.

The colors of the organs in this photo are not real. The colors are created by the computer that shows the image.

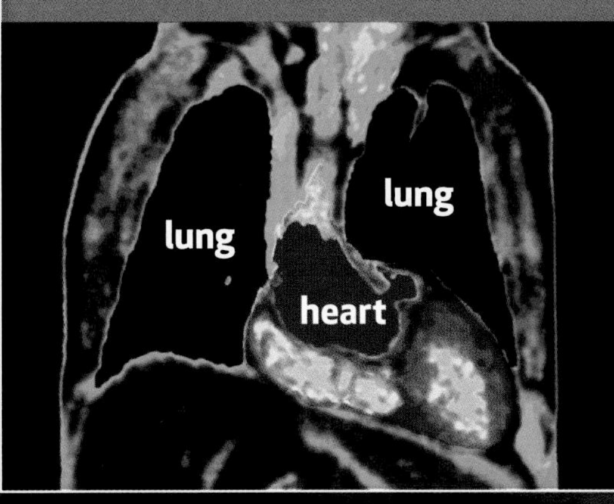

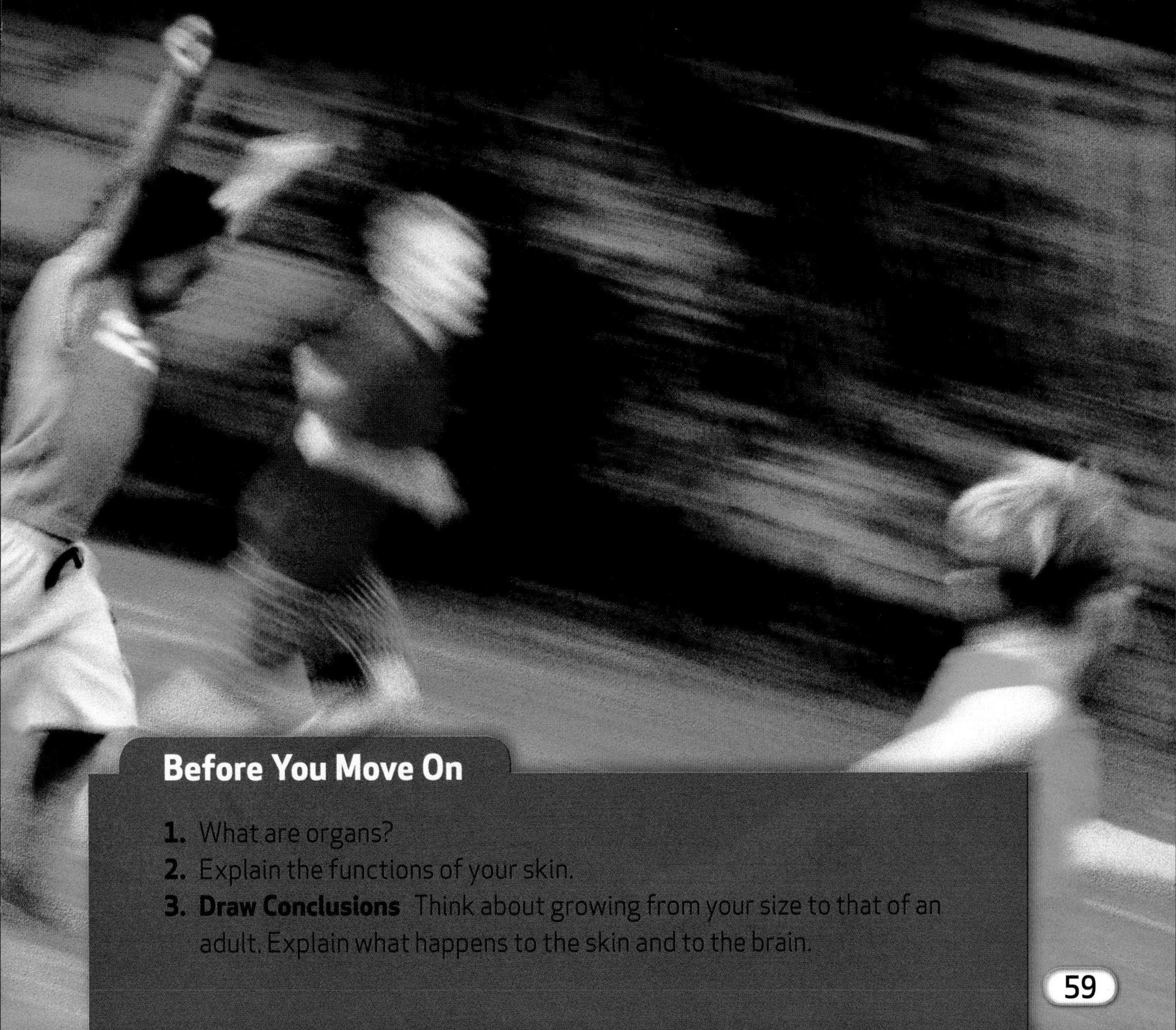

Before You Move On

1. What are organs?
2. Explain the functions of your skin.
3. **Draw Conclusions** Think about growing from your size to that of an adult. Explain what happens to the skin and to the brain.

Moving the Human Body

Bones Think about a tent. Its shape is determined by the frame inside it. Your body is somewhat like that. Inside your skin you have a framework of bones. Bones are organs.

This framework of bones is called a skeleton. In addition to providing a support structure, these organs do other jobs, too. They protect other organs, help make blood, and store minerals that your body needs.

You might think that bones are solid organs. But most are not. If you looked inside a bone such as the one in the upper leg bone, you would see it has a soft material in the center and lots of holes like a sponge near the ends. These different parts of a bone help it do its jobs.

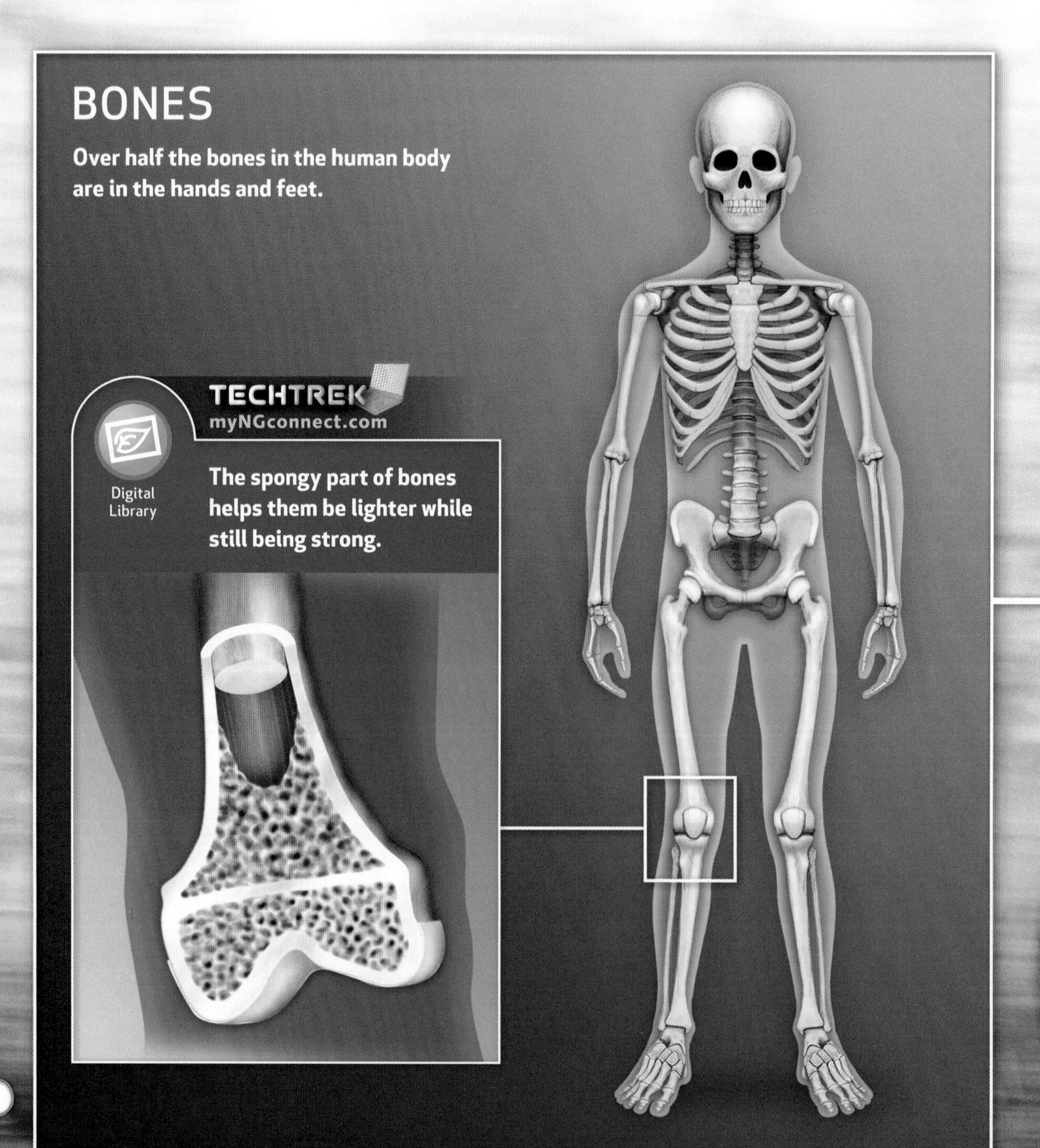

An adult human body has 206 bones of different sizes and shapes. Bones in your arms and legs are long and slender. Those in your hands and feet are short and stubby. Your skull is made of flat bones. The bones that make up your spine are rounded and fit together like a stack of puzzle pieces.

Bones come together at joints. Most of your joints are movable. Some joints, such as those in elbows and knees, bend like door hinges. Joints between chest bones allow slight back and forth movement, helping you breathe. The joints between the bones in your skull do not move at all. How do you think this is helpful?

The bony kneecap sits over the joint where the upper and lower bones of the leg come together.

Muscles The joints where bones connect determine where your body can bend, but bones cannot move on their own. Your muscles help do that job. When any part of your body moves, inside or out, muscles are at work. Muscles contract, or pull. When muscles relax, they stretch out.

Three kinds of muscles make all the movements in your body. Skeletal muscles are connected to bones. When a skeletal muscle contracts, it pulls on a bone and causes movement at a joint. When you flex your arm or pick up something heavy, you can see some of your arm muscles bulge. Movement of your skeletal muscles is voluntary—that means you decide to move them.

This boy is using many of his estimated 600 skeletal muscles in the body to stand straight, hold his skis over his head, and smile!

But not all of your muscles are connected to bones. Organs such as the stomach, bladder, and blood vessels contain smooth muscle tissue. Smooth muscles are involuntary. That's good because it means you don't have to think about having your stomach mix up your food!

Your heart is the only organ made of cardiac muscle, which is also involuntary. It contracts and relaxes, pumping blood throughout your body your entire life. It only rests between beats!

MUSCLES

Compare the action of the muscles in both diagrams.

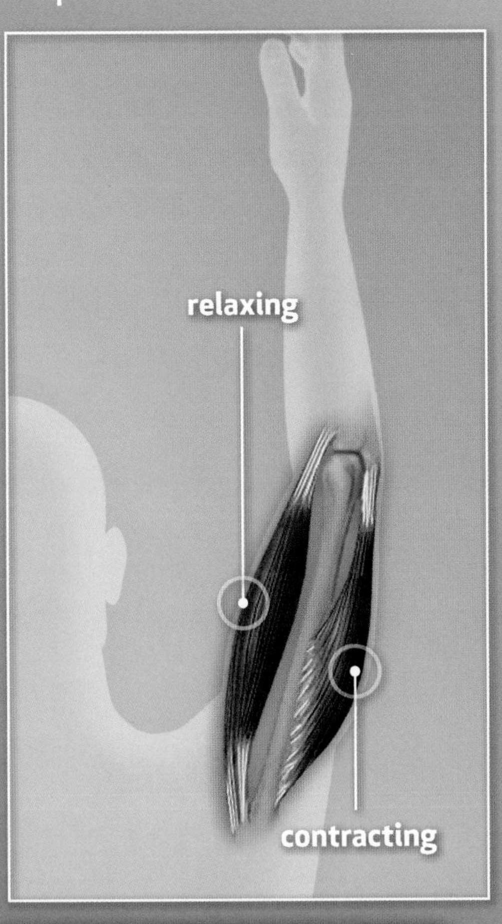

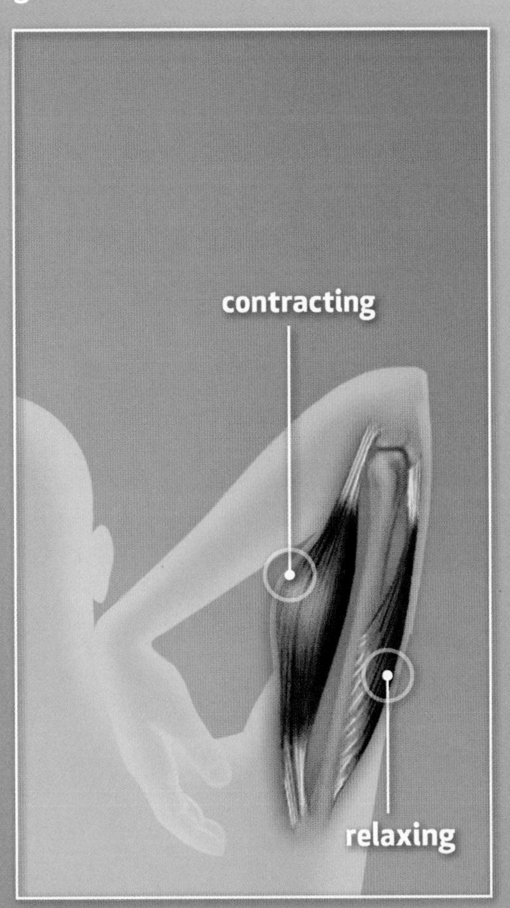

Before You Move On

1. What do bones do?
2. What do muscles do?
3. **Generalize** Bend your finger. Tell what is happening to the muscles in your finger.

Moving Blood and Oxygen

Blood Adults have about 5 liters (about 10 pints) of blood moving through their bodies. What does it do? Where does it go? How does it keep moving?

Blood delivers nutrients and oxygen to all parts of the body and carries away wastes. Blood moves through a network of vessels made of smooth muscle. Blood circulates in blood vessels throughout your body all of the time, even while you sleep.

An **artery** carries blood away from the heart. Arteries have muscle in their walls, which helps push the blood throughout the body.

Arteries connect to capillaries where the exchange of materials between the body and the blood occurs. Capillaries are the tiniest blood vessels. Capillary walls are so thin that oxygen moves out of the blood, through the wall of the capillary, and into the body. Waste products move from the body back into the blood.

Capillaries lead to **veins**. These blood vessels carry blood back to the heart. Then the blood circulates again.

The goalie's blood is moving quickly, carrying food and oxygen to all his muscles.

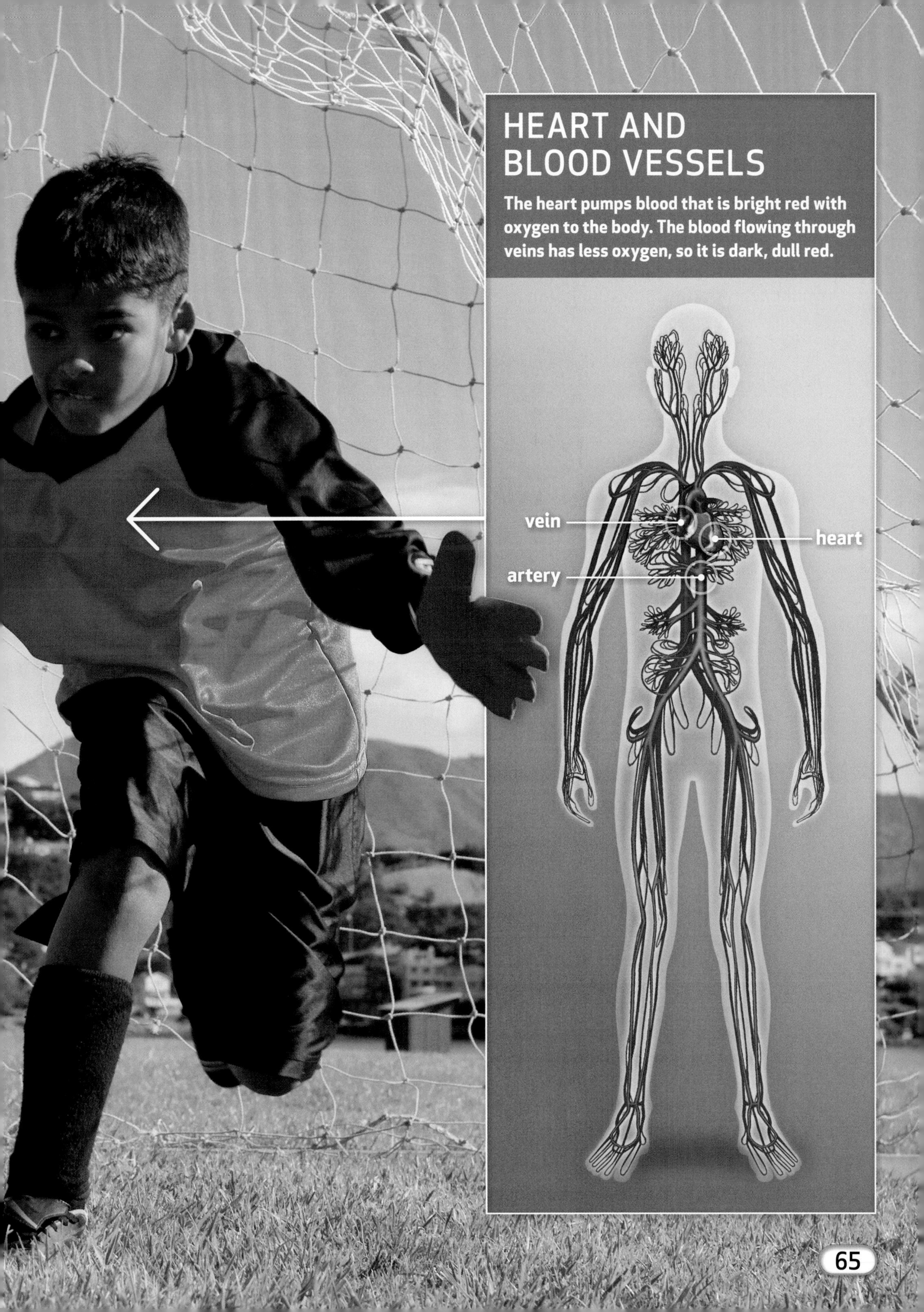

HEART AND BLOOD VESSELS
The heart pumps blood that is bright red with oxygen to the body. The blood flowing through veins has less oxygen, so it is dark, dull red.
vein
heart
artery

The Heart How does your blood keep moving through your blood vessels? It is pumped, nonstop, by your heart. The cardiac muscle contracts and relaxes, over and over. When the heart contracts, it pushes blood out to the body. When the heart relaxes, it fills with blood again.

The heart has four chambers, or sections, that are stacked, two above and two below. Blood from the body enters the top right chamber and then flows to the lower right chamber. This chamber pumps the blood to the lungs. Then the blood enters the top left chamber and flows to the lower left chamber. This chamber pumps blood out to the rest of the body. That takes a big push! The lower left chamber of your heart has the thickest muscle because it pumps so hard.

Your heart pumps faster or slower depending on what activity you are doing. For example, it pumps slower when you are sleeping. It pumps faster when you are running.

Science in a Snap! Feel the Beat

Lightly place your first two fingers of one hand over the inside of the wrist on your other arm. Move your fingers around slowly. Stop when you feel a throbbing. The throbbing is your pulse.

Count the number of pulses in 10 seconds. Multiply that number by 6. This is the approximate number of times your heart beats in one minute.

Would this number be smaller or larger after you jumped rope for 5 minutes?

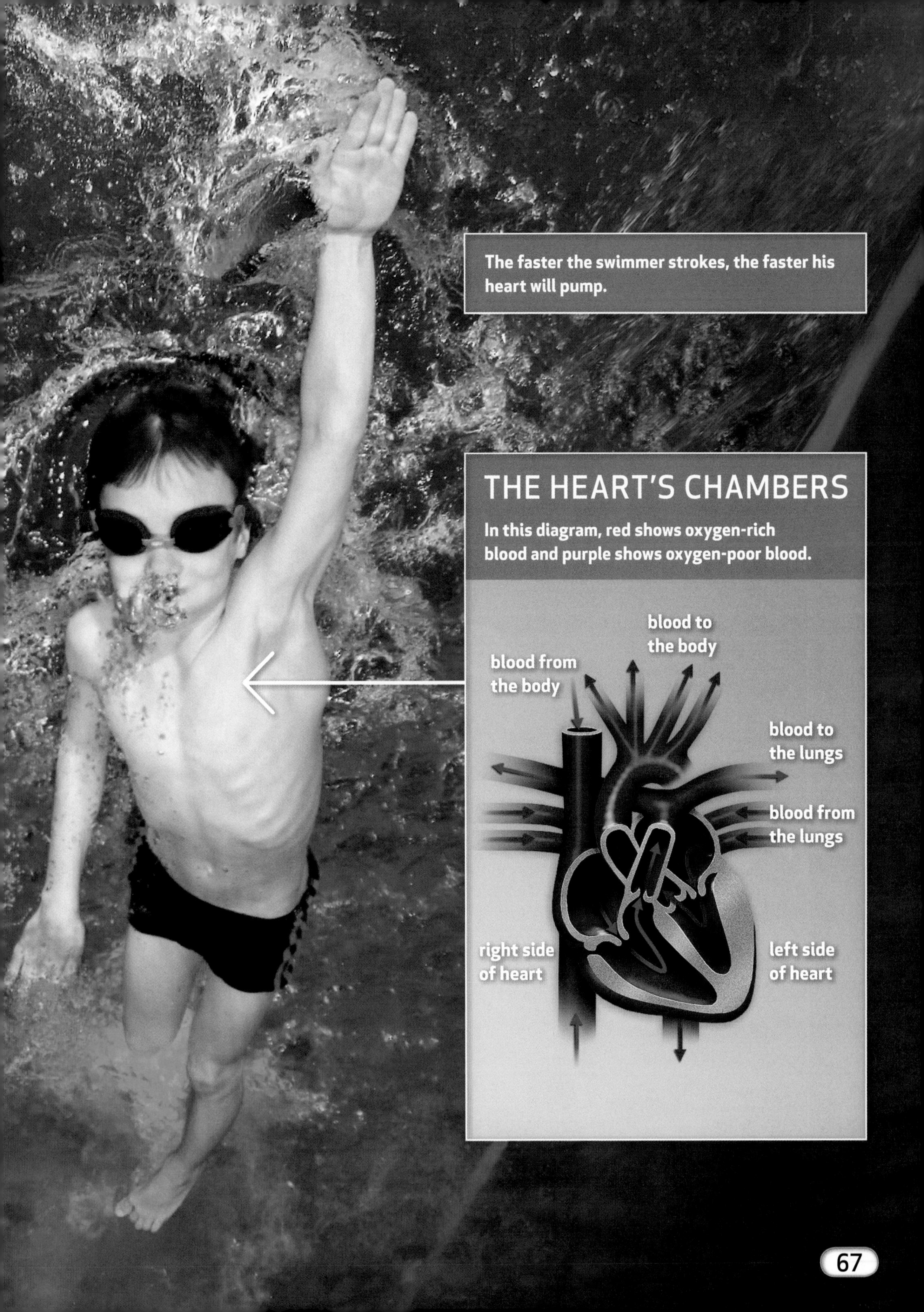
The faster the swimmer strokes, the faster his heart will pump.
THE HEART'S CHAMBERS
In this diagram, red shows oxygen-rich blood and purple shows oxygen-poor blood.
blood to the body
blood from the body
blood to the lungs
blood from the lungs
right side of heart
left side of heart

Lungs When you inhale, or breathe in, you take air into your body. This allows your blood to get the oxygen from the air that your body needs. Your body uses that oxygen and produces carbon dioxide as waste. When you exhale, or breathe out, you push out the carbon dioxide waste.

Your diaphragm is a sheet of muscle beneath your lungs that helps you inhale and exhale. When the diaphragm contracts, it moves downward. Your ribs move out, and your chest area gets a little bigger. Your lungs fill with air. When the diaphragm expands, it relaxes and moves back up. Your ribs move back in, and your chest area gets a little smaller. You exhale, and your lungs empty.

Just as your involuntary heartbeat continues night and day, your breathing continues all of the time. You can control your breathing for short periods, but when you stop thinking about it, your breathing continues as an involuntary function, even when you are asleep.

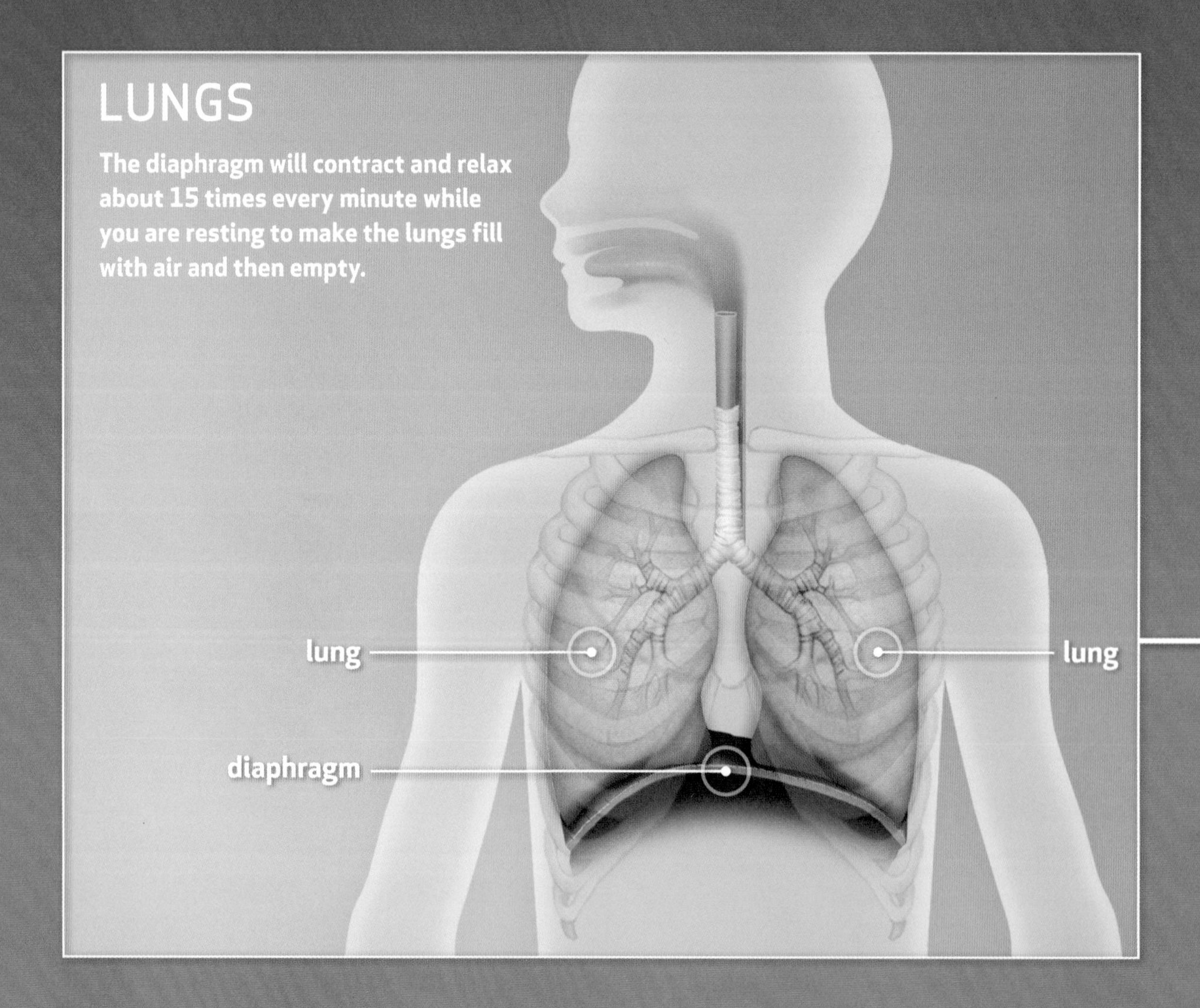

LUNGS

The diaphragm will contract and relax about 15 times every minute while you are resting to make the lungs fill with air and then empty.

You breathe air into your lungs, which are the main organs for breathing. They are like inflatable bags full of tubes that branch out like tiny trees. These tubes are the places where blood and air meet and exchange gases.

You breathe harder when you run because your muscles are working harder. They need more oxygen to do their job. When this happens, your heart beats faster to move more blood through your lungs. Your lungs must bring in oxygen and get rid of carbon dioxide more quickly.

This girl's balloon contains more carbon dioxide than the air around it.

Before You Move On

1. How does blood move through the body?
2. Explain how the heart and lungs work together.
3. **Predict** Will your heart pump and your diaphragm contract more or fewer times each minute after you go to sleep tonight than they did while you were eating dinner? Explain.

Getting Food and Removing Wastes

Are you hungry? If so, your body is telling you it needs food. But your body cannot use the food you eat until it is digested, or broken down into nutrients the body can use.

Digesting food begins when you chew it. Chewing breaks food into smaller pieces and mixes the food with saliva. This is just one of the juices your body makes that helps in digestion. Saliva begins to digest some foods. When you swallow, the crushed food travels through a tube to your stomach.

The stomach is a baglike organ where food mixes with water and other juices your body makes. The muscles in the stomach cause it to churn the food so it is mixed well with the juices. The soupy mixture then moves into the small intestine. This organ is a very long, winding tube. Most digestion happens here.

Two organs that help digestion are the pancreas and the liver. Both organs make certain juices that are dumped into the small intestine. The pancreas makes juices that help digest proteins and starches. The pancreas also helps keep the right amount of sugar in your blood. The liver produces bile, which helps digest fatty foods.

As food moves through the small intestine, the nutrients in food are absorbed for the body to use. Anything that cannot be digested moves into the large intestine. Here, water is absorbed into the blood. The waste material stays in the large intestine until it leaves the body.

TECHTREK myNGconnect.com

Enrichment Activities

ORGANS IN DIGESTION

A hamburger takes about 24 hours for the organs to digest it.

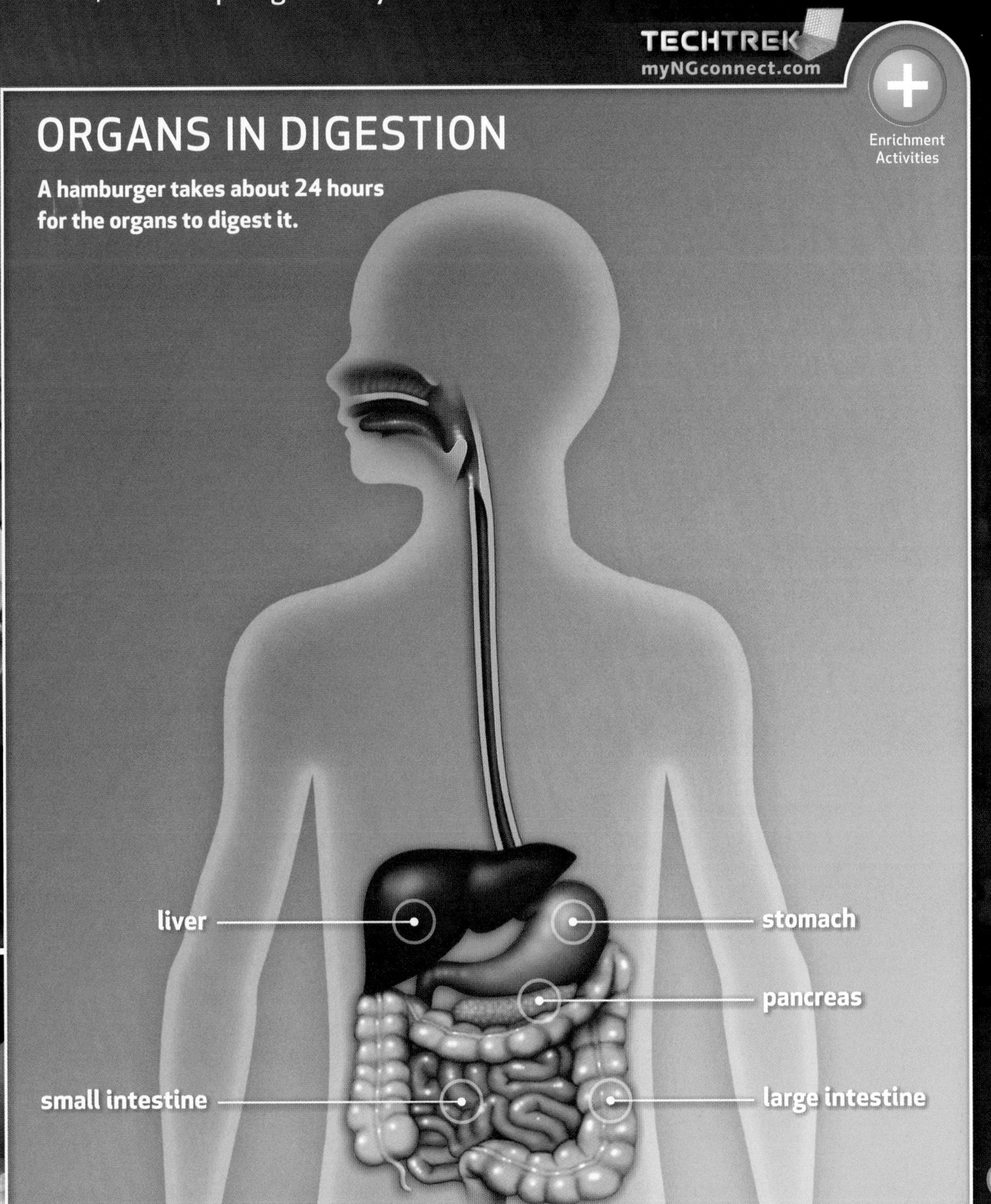

Kidneys and Bladder

Thinking, working, playing, growing, sleeping, or doing anything requires energy. You might eat a peanut butter sandwich to get energy to do something. As your body uses the nutrients from the peanut butter and bread, the body makes other materials your body cannot use. The body has to get rid of these waste materials. The body gets rid of materials it can't use much like you take out materials you can't use for garbage and recycling.

The job of some organs is to get rid of wastes.

The body's wastes move into the blood through the walls of the capillaries. As blood circulates throughout the body, it is filtered by not only the liver but also the **kidneys**. The kidneys are organs that remove wastes and extra water from the blood. If you followed a single drop of blood as it travels throughout the body, you would see that it moves through the liver and kidneys each time.

Your kidneys mix the wastes they filter from the blood with the water that is absorbed by the large intestine. This mixture is urine. The urine travels through long tubes to the bladder. Your bladder collects and stores the urine until it leaves your body.

KIDNEYS AND BLADDER

The kidneys are about the size of your fist. They lead to the muscular bladder.

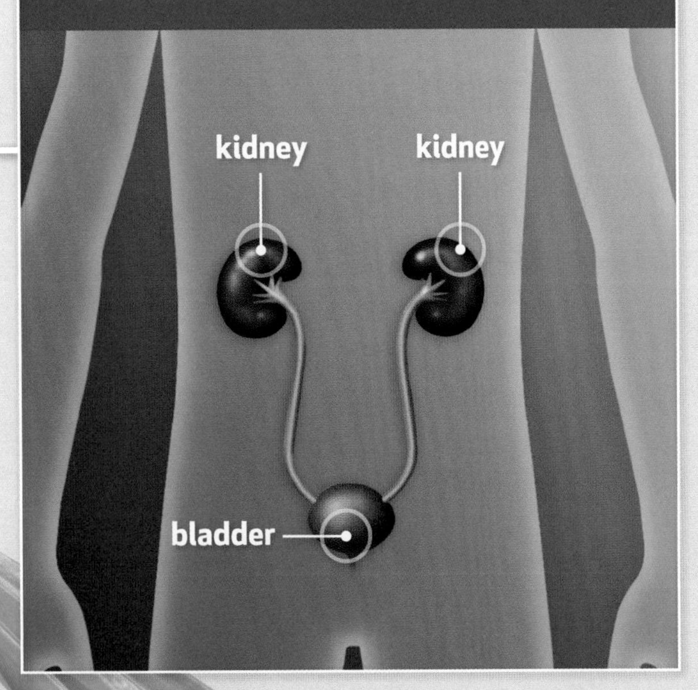

Before You Move On

1. Which organ holds food and mixes it with juices made by your body?
2. Which organs add juices to the small intestine to help digest food?
3. **Apply** Why is it important to drink water?

Controlling the Human Body

Your brain is the organ that controls your entire body. Your brain takes in information about the world around you and tells your body how to respond. Sometimes you are aware of this process. Sometimes you are not.

The three main parts of your brain control different functions. The cerebrum is the largest part of the brain. It controls your voluntary muscle movements, your speech, and your senses. The cerebrum also enables you to think. It stores your memories and enables you to learn.

The cerebellum is located at the back of the head. This part controls balance and posture. It also fine-tunes movements so that the body can move gracefully.

Your brain controls the actions of your body like this man controls the actions of the machine.

The brain stem controls body functions that are automatic, such as breathing, swallowing, and your heartbeat.

Your brain sends and receives information through your spinal cord. The spinal cord is a bundle of nerves. It directs signals from your brain to the other nerves that branch out into your body.

Nerves form a weblike network. They carry signals that allow you to sense the world around you and control your body. You do not always know about the messages that nerves carry. Nerves help your body parts do involuntary jobs, such as digesting food and circulating blood.

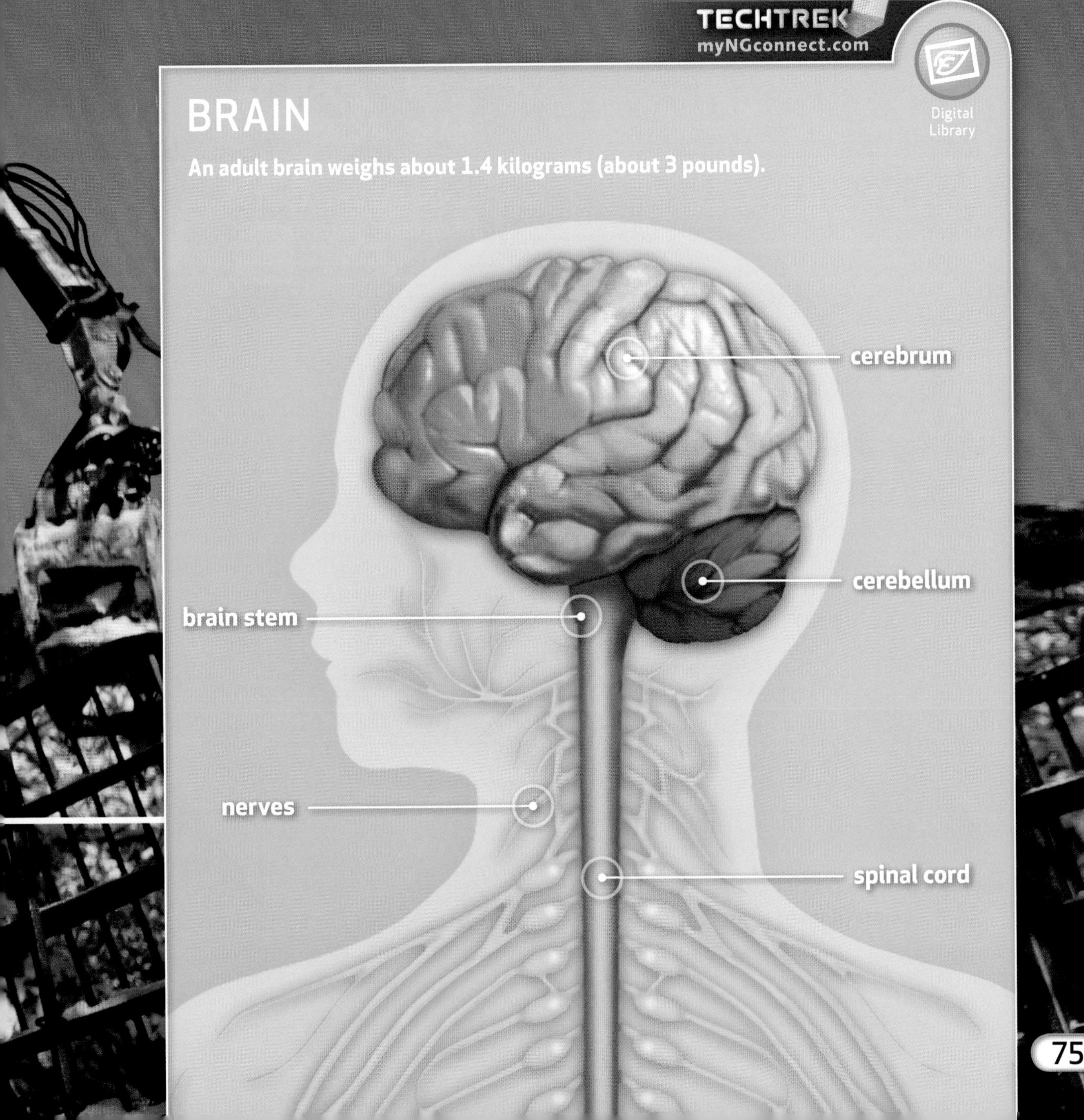

When you go to the movies, you can probably smell popcorn as soon as you walk into the theater! You see the action on the screen and hear the sound effects. As you eat your popcorn, you use your sense of taste. You use the sense of touch in your fingers to find and pick up the popcorn in the dark. You feel the cold of the lemonade you drink.

What is happening in these moviegoers' brains? They are using all of their sense organs at once. Sense organs are full of nerves that detect different kinds of information. They are the body's way of gathering information. You have sensory organs that deliver this information to your brain. Your brain uses information from your senses to determine action.

Scientists think the front part of your cerebrum controls your laughter!

SENSORY ORGANS

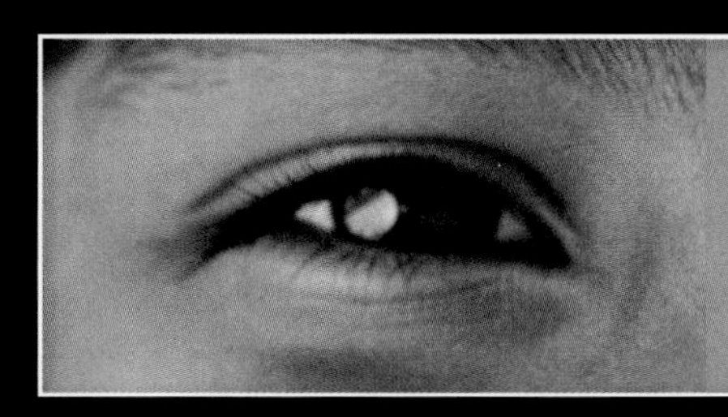

EYES **Nerves in the eye detect light and color. The signals travel to your brain, which tells you that you see an object or a scene.**

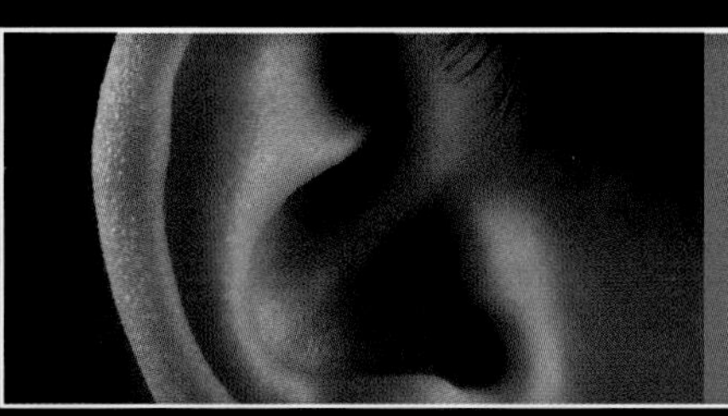

EARS **Vibrations in the air move into the ear. Nerves detect the vibrations and send signals to your brain, which tells you that you hear sound.**

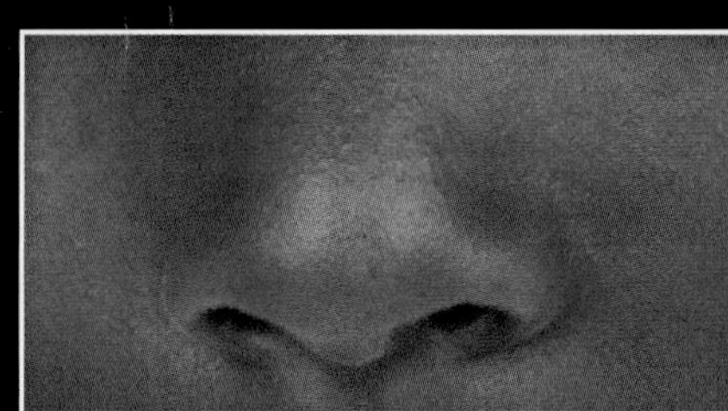

NOSE Very tiny particles in the air move into your nose. Nerves detect the particles and send signals to your brain, which tells you that you smell odors.

TONGUE **Nerves in the tongue detect four tastes—sweet, sour, bitter, and salty. Nerves send signals about combinations of these to your brain, which tells you that you are tasting different flavors.**

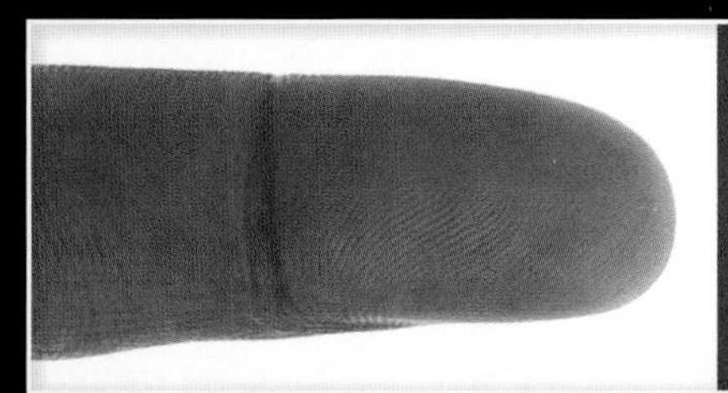

SKIN **Nerves in the skin detect heat, cold, contact, pressure, and pain. The nerves send signals to your brain, which tells you what you feel.**

Before You Move On

1. Name the human sensory organs.
2. Describe voluntary and involuntary ways that your brain is controlling your body right now.
3. **Analyze** How does your brain interact with your body's other organs?

Organs in Plants and Animals

Plants and animals other than humans have organs, too. The organs in those living things can carry out jobs similar to the jobs organs do in humans.

Remember that your body is supported and protected by your skeleton. Dogs, lions, goldfish, alligators, and eagles all have skeletons on the insides of their bodies too. Like yours, their skeletons support their bodies and protect their internal organs.

What organs are enabling both the boy and the dog to run?

Other kinds of animals have hard outer coverings that support and protect them. These animals have exoskeletons. Some animals, such as clams, are enclosed completely by hard shells. The exoskeletons of lobsters and crabs form layers like armor on their outer bodies. Insects have exoskeletons, too.

Plants also have an organ for support—the stem of the plant. Stems hold the plant upright. They also move materials between the roots and the leaves. Stems can be very short or as tall as the trunk of a giant sequoia.

The exoskeleton of this insect is shiny.

Plant stems hold leaves up to the sunlight.

Other Functions Just like you, other living things also must have ways to get the materials they need to live and grow. For example, animals have organs that digest food and get oxygen. You can see in the charts how their organs compare to those in humans.

GETTING **FOOD**

COWS Cows eat mostly tough plant material. The cow's four-sectioned stomach helps break down the grasses into nutrients the cow's body can use.

CARDINALS Birds have a two-sectioned stomach. One part is called the gizzard, which breaks down seeds. The gizzard is very muscular and helps grind up food. Because birds do not have teeth, they do not chew their food.

SPIDERS Many spiders and flies begin to digest their food before they actually eat it! The spider puts its digestive juices on or into solid food to turn it into a liquid. Then they drink it.

GETTING **OXYGEN**

RABBITS All mammals, such as rabbits, have lungs that are structured much like human lungs and do the same thing. All birds and reptiles have lungs, too.

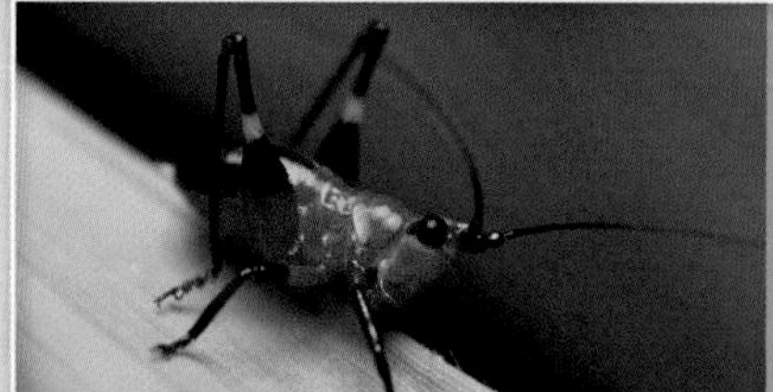

GRASSHOPPERS Insects, including grasshoppers, have networks of tubes that carry gases to different parts of their bodies. Spiders have these tubes, too.

HUMPHEAD CICHLIDS Fish, such as cichlids, have gills. Water moves in through the fish's mouth and out over the gills. Oxygen moves from the water into the fish's blood.

Plants also need food and air, but they get and use materials differently from animals. Plants cannot move around and eat food. Instead, most plants make their own food. Leaves are organs with parts that can make food using light, water, and carbon dioxide. Leaves take in the carbon dioxide and oxygen from the air that the plant needs.

Roots are another important organ in plants. This organ anchors the plant in the soil. It also takes in water from the soil that the plant uses. Water and food move between the leaves and roots through the stem.

The organs of a plant carry out the plant's life processes.

Before You Move On

1. Identify structures that provide support in animals and in plants.
2. How do plants and animals get the energy they need?
3. **Generalize** What do all plants and animals have in common?

MAKING SENSE OF SENSES

Animals use their senses to keep track of their surroundings—to find food and to avoid threats. Many animals' senses are similar to your own. But their sensory organs can be very different from yours. The sharpness of animals' senses is different from yours, too.

Elephants have very sharp senses of hearing and smell. But their eyesight is not as good as a human's.

Hearing Elephants and bats both have much more sensitive ears than humans have, but in different ways. Elephants can detect sounds that are too low-pitched for humans to hear. They use these low-pitched sounds to communicate with one another over very long distances. Bats can detect sounds that are too high-pitched for humans to hear. They use sounds that people cannot hear to locate insects for food.

Touch Some animals detect objects using the sense of touch—but without actually touching the objects. A cat's whiskers detect changes in air movements near an object. A seal's whiskers work the same way, but they detect movements in water. Insects, crabs, and lobsters have antennae that feel both objects and motion.

The big brown bat tracks insects using sound instead of sight.

Whiskers help cats feel their way in very low light.

Taste Insects use their antennae for more than touch. Many insects can also taste with their antennae. Catfish have antennae-like structures called barbels. Barbels are covered with taste sensors. Catfish do not see well, so they taste their way along the bottoms of rivers and lakes. Some types of blind fish have taste sensors covering most of their bodies. The bodies of earthworms are covered with taste sensors, too. You only taste with your tongue, but an octopus tastes with its tentacles. A housefly tastes with its feet!

Smell Those same insect antennae that taste and touch can also sense odors. Many animals have a much stronger sense of smell than humans. Some breeds of dogs can detect where a human simply walked a few days ago. You might be able to smell your lunch from across the room, but a grizzly bear can smell food buried deep in the ground. Some bears can detect scents in the air coming from over 24 kilometers (about 15 miles) away!

antenna

Catfish barbels have many more taste sensors than your tongue.

Insects use antennas to sense touch, taste, and smell.

Sight You have two eyes in the front of your head, but you can turn your head to get a better look at something. A spider does not have a neck on which to turn its head, but it can still look everywhere around it. Spiders have as many as eight eyes bunched near the top of the head. Crabs have two eyes like you, but they are perched on top of stalks. Eagles, hawks, and owls have sharp eyesight and can spot small prey from high in the air. Animals that spend their entire lives in dark caves or beneath deep water might not have eyes of any sort.

A wolf spider has two eyes on the top sides of its head plus the six looking at you!

The fiddler crab's eyes are perched on stalks.

Conclusion

Organs are structures that carry out specific jobs in humans. Bones provide a framework for the body and work with muscles to move the body. The heart pumps blood through blood vessels to the lungs and the rest of the body to get oxygen and take away wastes. The liver and pancreas work with the stomach and intestines to break down food into a form the body can use. Kidneys and the bladder remove wastes. The brain and sense organs work together to sense the environment. Other animals and plants also have organs that carry out many of these same jobs.

Big Idea Organs do specific jobs in living things as they live and grow.

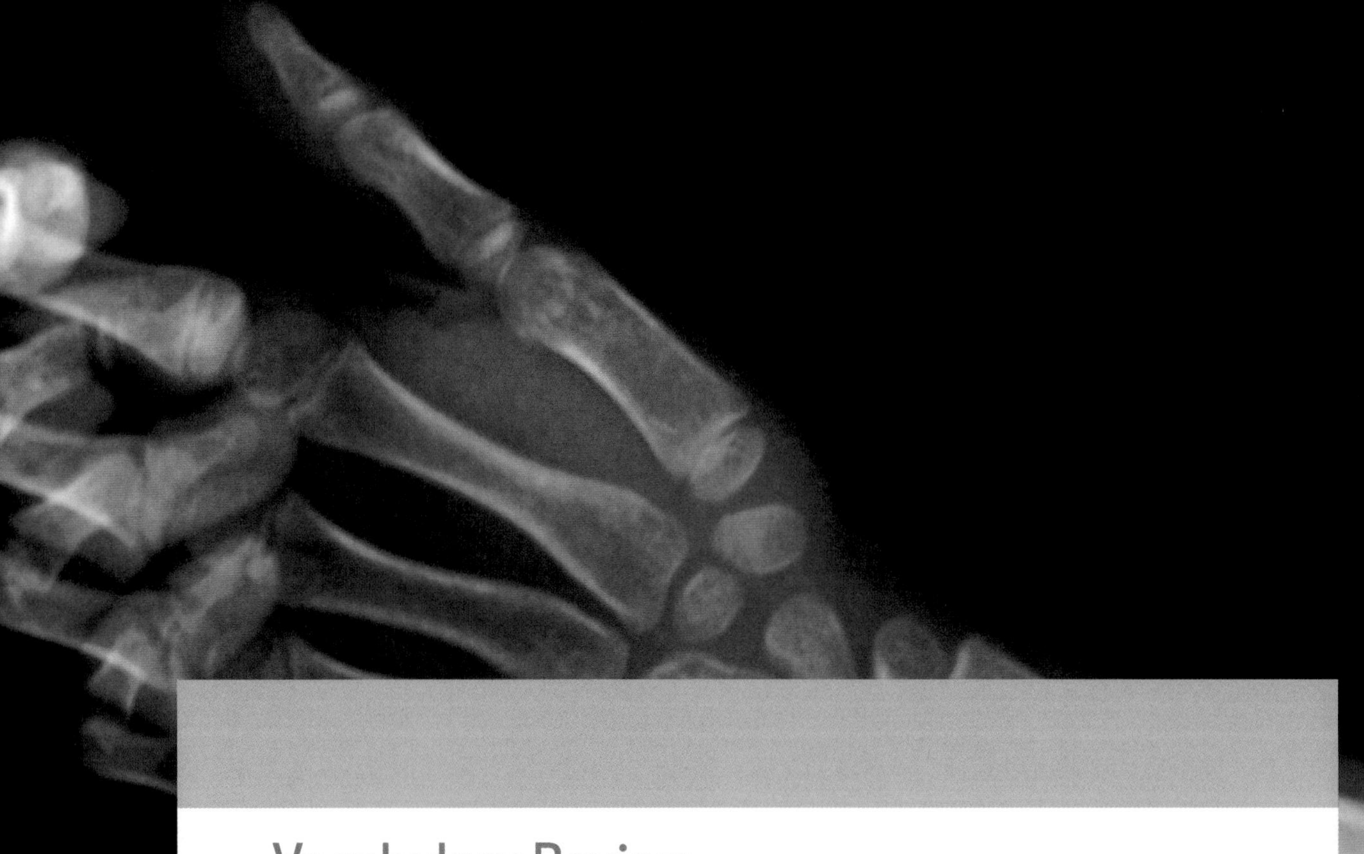

Vocabulary Review

Match each of the following terms with the correct definition.

A. pancreas
B. organ
C. artery
D. liver
E. vein
F. kidney

1. An organ that produces bile to help digest fatty foods
2. An organ that removes wastes and extra water from the blood
3. An organ that helps digest proteins and starches
4. A blood vessel that carries blood toward the heart
5. A blood vessel that carries blood away from the heart
6. A structure that carries out a specific job in the body

Big Idea Review

1. **List** Choose five organs you learned about and tell what they do.
2. **Describe** What are two organs that help filter the blood?
3. **Cause and Effect** How do bones and muscles work together when you raise your arm?
4. **Sequence** Trace a drop of blood as it leaves the heart and circles back again.
5. **Generalize** How do organs in animals and plants compare?
6. **Evaluate** Do you think the heart or brain is the control center of the body? Support your answer.

Write About Organs Working Together

Explain What is happening in this photo? What organs are working together to make this action possible?

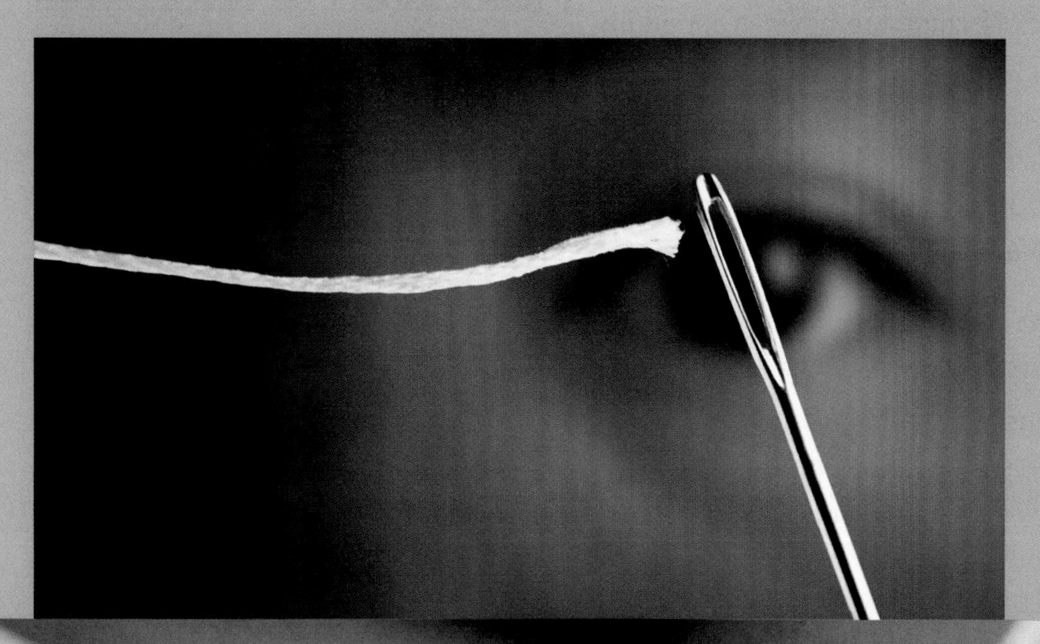

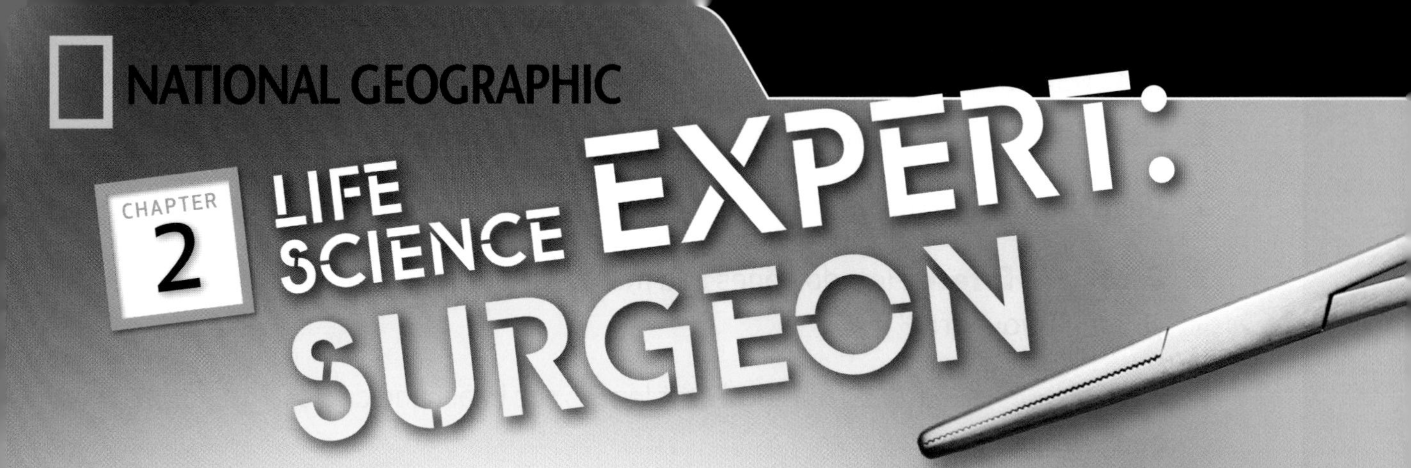

LIFE SCIENCE EXPERT: SURGEON

Dr. Lori Arviso Alvord is a Navajo surgeon. She received her training at Stanford Medical School. She combines traditional Navajo healing with her training to treat patients.

What do you do as a surgeon?

I work on the inside of people's bodies and fix problems they are having. For example, I might remove an appendix or a gallbladder if it is causing problems. I am trained to do many different operations on the organs of the body. Sometimes, when I operate, it saves a person's life.

What do you remember liking about science when you were in school?

Science is all about discovery and understanding the world. I loved the study of animals and how they live in their environment. I also enjoy any kind of chemistry experiment that makes something cool happen!

What's a typical day like for you?

I spend most days either in the clinic or in the operating room. On clinic days, we start at around 8 a.m., end around 5 p.m., and see patients all day.

Dr. Arviso Alvord works closely with people in all parts of her job.

We figure out what sort of surgical problem someone has, and then often schedule surgery. Sometimes we are called to the emergency room to see a patient who needs surgery right away. On an operating room day, we spend all day operating. We might do three to five operations. We are part of a team who work closely together all day long.

What has been the coolest part of your job?

The first really cool part is doing an operation. You get to work with your hands. You do a lot of dissection and clamping and tying of blood vessels. Sometimes you operate using instruments through small openings in the skin, called laparoscopic surgery. This is both fun and interesting.

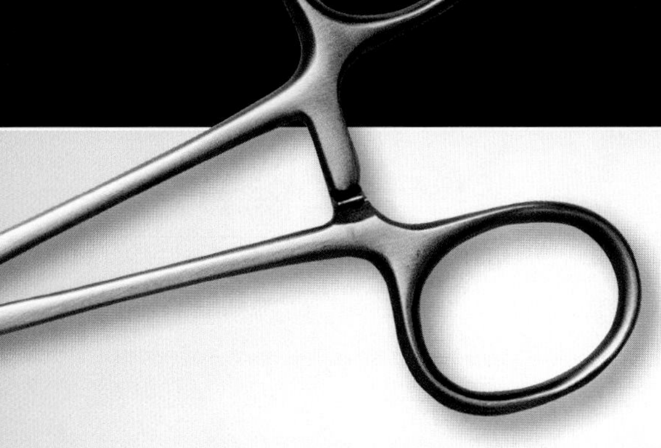

The second cool thing is that you do something that makes a huge difference in someone's life. It is an honor to be able to help people so very much.

What has been your greatest accomplishment so far?

My greatest accomplishment so far has been my ability, as a surgeon, to make a difference in people's lives, to relieve their pain, and to help them to get well. Sometimes I can save a life. But I've also had success as an author and a public speaker. I have served as a role model for many Native American children. I hope I have shown them that they can do whatever they set their minds to do. I also hope that my work to create "healing environments" for patients, by using Native American philosophies of healing, has helped to move medicine in a direction that helps all patients.

TECHTREK
myNGconnect.com

Dr. Arviso Alvord works with others during surgery.

Digital Library

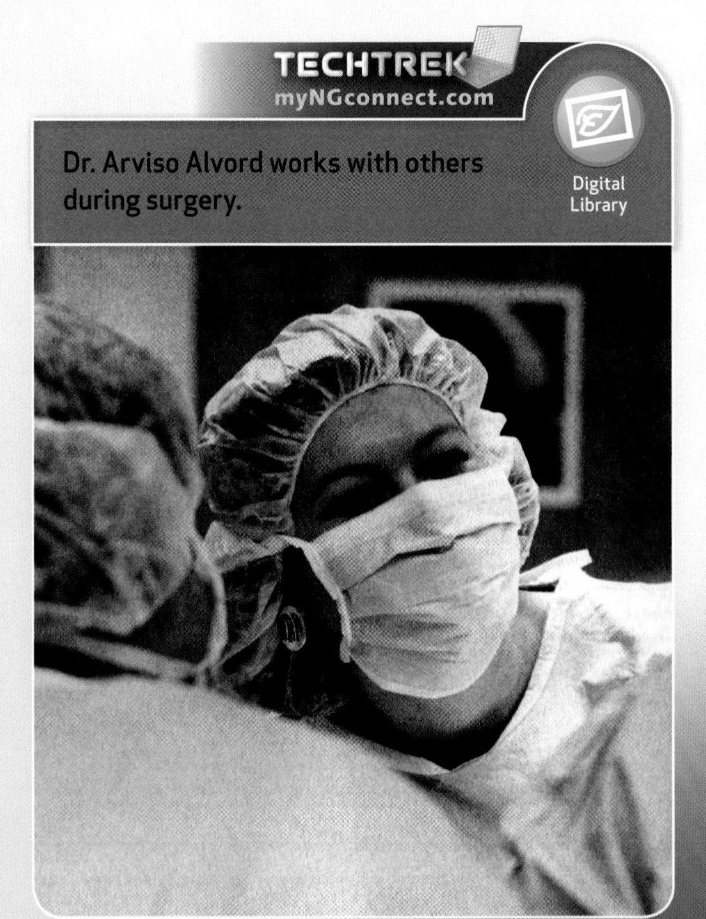

What advice would you give young people who want to become surgeons?

Learn all you can about medicine and surgery and be sure that's what you want to do. Then do things that develop your mind, such as music, art, reading and literature, athletics and dance. Also, since medicine is all about helping people, develop the right attitude toward people. You can start by getting involved in community service projects.

NATIONAL GEOGRAPHIC

BECOME AN EXPERT

Ironman Triathlon Race: Body Organs Working Together

A triathlon is a race of three parts: swimming, bicycling, and running. Triathlons are different lengths, but the hardest is the Ironman Triathlon. Racers have to swim 3.8 km (2.4 mi), bike for 180 km (112 mi), and then run for 42.2 km (26 mi).

Racers have to train incredibly hard to prepare their bodies for such a long race. By the last part of the race, their bodies are already exhausted even if all factors, such as the weather, are perfect.

Hard work takes place in the racer's body. The heart pumps blood; the lungs take in oxygen; muscles pull on bones to move the body forward. These **organs** and many others work especially hard during a race.

Racers start the first part of the race—swimming.

organ

An **organ** is a structure that carries out a specific job in the body.

Energy for the Body

Triathletes need a great deal of energy for over twelve hours of racing. The energy comes from food.

Digestive organs turn food into nutrients the body can use. Digestion begins in the mouth, where teeth chew food and mix it with saliva. Food moves to the stomach and digestive juices are added. Then the food becomes a thick liquid and moves into the small intestine, where digestion is finished. Juices from the **pancreas** and the **liver** help break food down into nutrients. The small intestine absorbs the nutrients into the blood. Any remaining food passes to the large intestine, where water is absorbed.

pancreas
The **pancreas** is an organ that helps digest proteins and starches.

liver
The **liver** is an organ that produces bile to help digest fatty foods.

Getting Food and Oxygen

Training strengthens organs, such as the heart, for high performance. The heart muscle must be strong enough to pump blood—carrying nutrients and removing waste products at the same time—throughout the grueling race.

Blood travels in a circle-like path that has no beginning and no end. If you start on the left side of the heart, blood rich in oxygen moves through **arteries** to the body. Blood reaches the capillaries, where oxygen moves out of the capillaries into the body. Carbon dioxide then moves back into the blood. Food and waste products are also exchanged. Then, through the **veins**, the blood moves back to the right side of the heart.

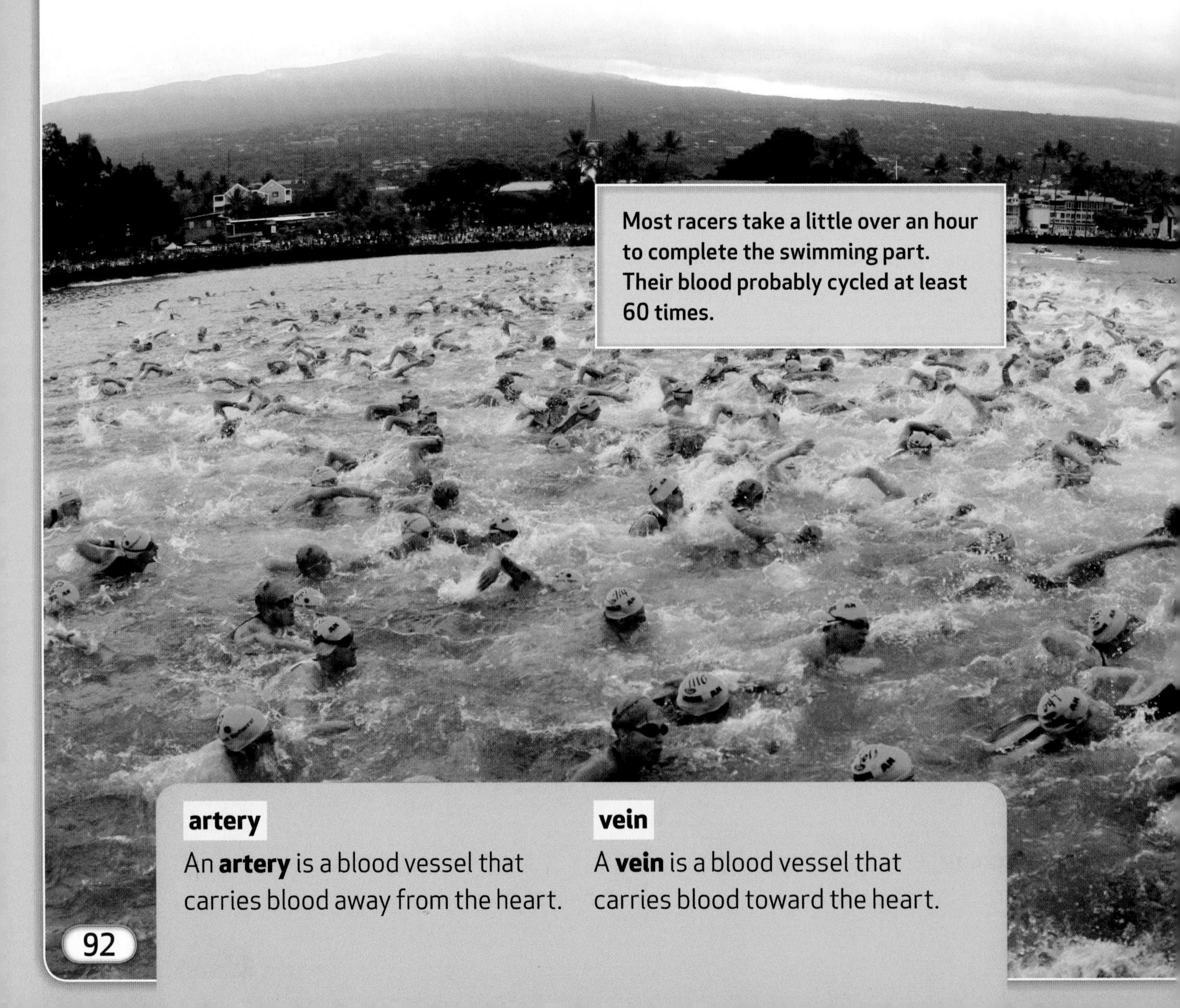

Most racers take a little over an hour to complete the swimming part. Their blood probably cycled at least 60 times.

artery

An **artery** is a blood vessel that carries blood away from the heart.

vein

A **vein** is a blood vessel that carries blood toward the heart.

From the right side of the heart, blood travels to the lungs. Here the carbon dioxide is exchanged for oxygen, and the blood moves back to the left side of the heart. The cycle repeats hundreds of times during the race.

The racer has to inhale and exhale just right so only air enters the lungs. Breathing starts at the nose and mouth. The racer inhales air. A muscle called the diaphragm, at the base of the rib cage, flattens and lets the lungs expand when the athlete inhales. The lungs fill with air. When the racer exhales, the diaphragm relaxes and the lungs empty.

This racer exhales underwater.

Support and Movement

Just like the bicycles they are riding, racers need strong frames. The long bones and muscles of the legs are worked extra hard during the bicycling part of the race.

TECHTREK
myNGconnect.com

Digital Library

The bicycling part takes about six hours for most racers. Think about how many times a thigh muscle would contract during that time!

The racer's skeleton flexes and bends at its joints to pedal the bicycle. The many small bones of the backbone enable the bicyclist to bend over smoothly to reach the handlebars. The elbows bend slightly. And the joints between the many bones in the wrists and hands are always flexing to ensure the right grip.

The skeletal muscles, or those muscles attached to bones, are working continuously during all parts of the race. Some muscles are used more in one part of the race than in others. Bicycling uses the muscles of the legs and lower body. Even the back muscles are contracting to keep the racer hunched down over the bicycle. Staying hunched over as much as possible means the racer is more streamlined and can ride faster.

Other kinds of muscles are at work, too. The cardiac muscle of the heart is contracting at a fast pace. The smooth muscles of the arteries are helping get the blood to the body. The diaphragm moves up and down in a regular pattern to get enough air into the body.

HOW BONES AND MUSCLES WORK TOGETHER

Muscles contract, or pull. Muscles often work in pairs to move bones. Notice which muscle contracts and which relaxes to pedal a bicycle.

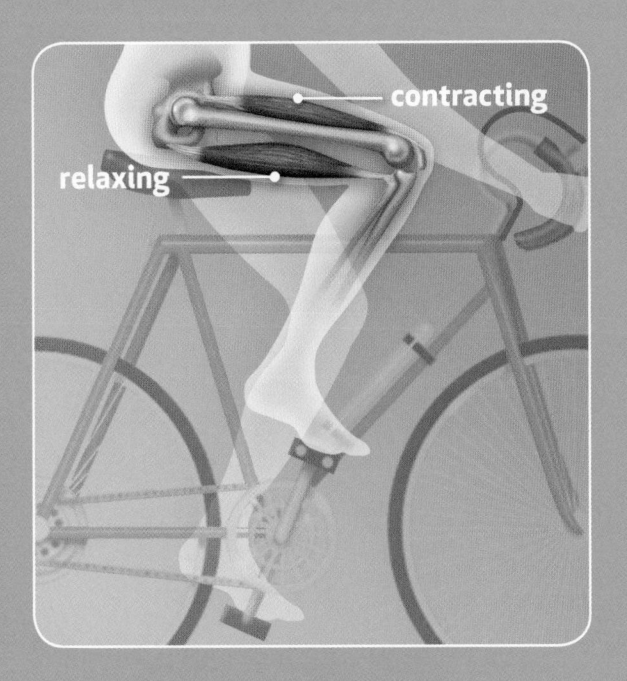

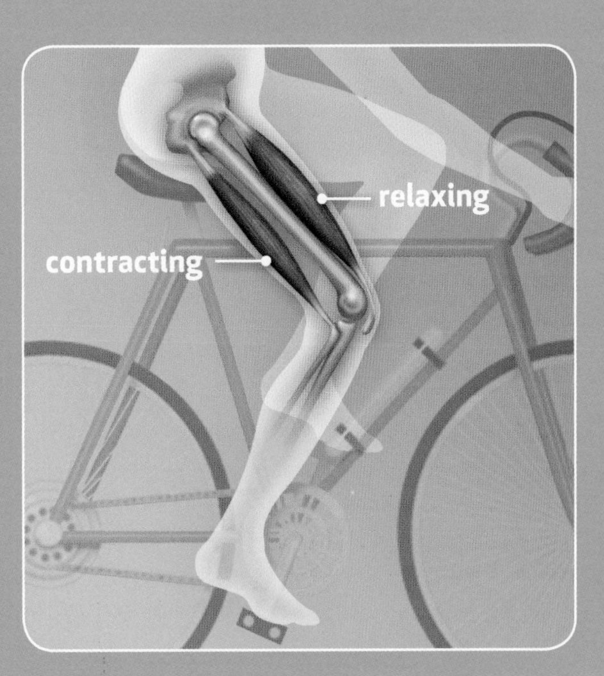

Control and Senses The work of the racer's body must be coordinated and controlled. The brain, spinal cord, and nerves command the functions of the entire body. They move messages extremely quickly to and from all body parts.

The nerves form a web throughout the body. They carry messages to the spinal cord. From there the messages speed to the brain. The brain processes the information and responds. It sends orders back through the spinal cord to the nerves.

Running might seem somewhat automatic. But after the swimming and bicycling, it is not easy to continue on foot. The racer must be able to focus the brain on not tripping, especially if there's a small dip in the pavement. And that's in addition to what the brain is controlling already!

Most racers take about four and one-half hours to complete the running part. It can take longer when the weather is hot.

The body and brain respond to the environment. The racer gathers information about the temperature of water and air, the motion of waves, noise, and the location of holes along the side of the road. Sensory organs—the eyes, ears, nose, tongue, and skin—collect this information for the brain.

The skin does several other jobs during the race. One of the most important is keeping the racer's body at the right temperature. It also helps protect the body if the racer falls.

Removing Waste A racer's body produces waste as it works. All of the organs need constant cleaning. The liver and **kidneys** filter waste from blood. The two kidneys filter all the body's blood about two times each hour. They return clean blood to the blood vessels and mix the waste with water. The liquid waste, urine, goes to the bladder.

An exhausted racer barely crossed the finish line.

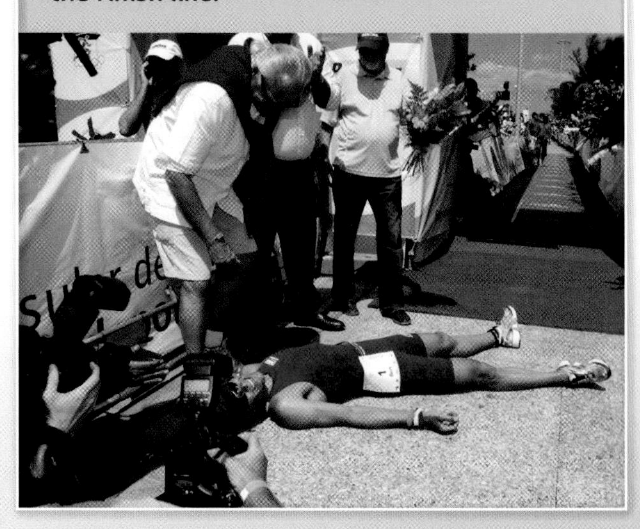

It has been many hours since the racers jumped into the ocean this morning.

kidney

A **kidney** is an organ that removes wastes and extra water from the blood.

Success! What a fatiguing race! The winner might finish in less than nine hours. Most racers take longer. But the organs in every racer have to be working their best. The triathlon pushes every organ to its limit. Strength training for organs and training them to work over long periods of time are very important for success. And each racer has to have a winning attitude!

Imagine being able to smile after winning this exhausting race!

NATIONAL GEOGRAPHIC

CHAPTER 2 SHARE AND COMPARE

Turn and Talk How do organs work in living things? Form a complete answer to this question together with a partner.

Read Select two pages in this section that are the most interesting to you. Practice reading the pages so that you can read them smoothly. Then read them aloud to a partner or small group. Talk about why the pages are interesting.

my SCIENCE notebook

Write Write a conclusion that tells the important ideas you have learned about how organs work in living things. State what you think is the Big Idea of this section. Share what you wrote with a classmate. Compare your conclusions. Did your classmate recall that muscles work in pairs to move bones?

my SCIENCE notebook

Draw Imagine what it is like racing in a triathlon. Draw a picture of an organ that helps a racer during a triathlon. Combine your drawing with some of your classmates in a triathlon racer's body outline. Present your group's racer by telling how the organs worked and if your racer won!

FLORIDA EARTH SCIENCE

What Is Earth Science?

Earth science investigates all aspects of our home planet from its changing surface, to its rocks, minerals, water, and other resources. It also includes the study of Earth's atmosphere, weather and climates. As Earth is an object in space, Earth science also includes the study of Earth's relationship with the sun, moon, and stars. People who study our planet are called earth scientists.

You will learn about these aspects of earth science in this unit:

WHAT MAKES UP THE SOLAR SYSTEM?

Our part of the universe, the solar system, includes many different kinds of objects. Revolving around the nearest star—the sun—you will find planets, moons, asteroids, dwarf planets, and comets. Earth scientists study these objects and Earth's relationship with them.

HOW ARE WEATHER AND THE WATER CYCLE CONNECTED?

Weather is the condition of the atmosphere around you. Water in the atmosphere is a part of weather. Earth's water is used over and over again as it cycles through a system called the water cycle. Through evaporation, condensation, and precipitation, water is constantly reused. Earth scientists study weather and the role of water in it.

MEET A SCIENTIST

Tim Samaras: Severe-Storm Researcher

Tim Samaras is a severe-storm researcher and a National Geographic Emerging Explorer. Tim studies tornadoes in order to give people better warning of future storms. He also studies tornadoes to find out more about when, where, and how they form.

Tim invented weather-measurement probes that record data such as humidity, air pressure, temperature, wind speed, and direction. Sometimes the probes have cameras inside that can record the inside of a tornado! Tim's goal is to place a weather probe in the path of a tornado. This can sometimes be dangerous work as he has to get close to a tornado. Tim uses probes and other weather devices to help predict when and where a tornado is going to hit.

About his work with tornadoes, Tim says, "Data from the probes helps us understand tornado dynamics and how they form. With this information, we can make more precise forecasts and ultimately give people earlier warnings."

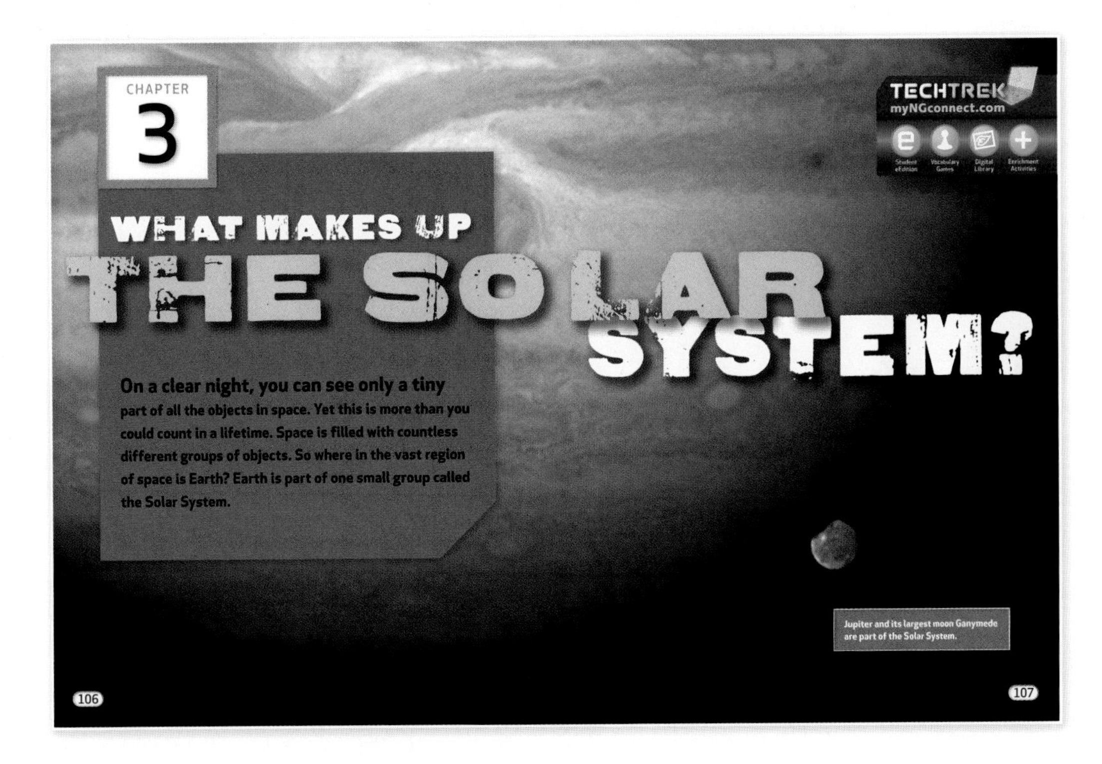

In Chapter 3, you will learn:

FLORIDA NEXT GENERATION SUNSHINE STATE STANDARDS

SC.5.E.5.1 Recognize that a galaxy consists of gas, dust, and many stars, including any objects orbiting the stars. Identify our home galaxy as the Milky Way. **STARS AND GALAXIES, THE SOLAR SYSTEM**

SC.5.E.5.2 Recognize the major common characteristics of all planets and compare/contrast the properties of inner and outer planets. **THE SOLAR SYSTEM, THE INNER PLANETS, THE OUTER PLANETS**

SC.5.E.5.3 Distinguish among the following objects of the Solar System, Sun, planets, moons, asteroids, comets, and identify Earth's position in it. **STARS AND GALAXIES, THE SOLAR SYSTEM, THE INNER PLANETS, THE OUTER PLANETS, OTHER OBJECTS IN THE SOLAR SYSTEM**

SC.5.E.5.2 Science in a Snap! Recognize the major common characteristics of all planets and compare/contrast the properties of inner and outer planets.

CHAPTER

3

WHAT MAKES UP THE SO

On a clear night, you can see only a tiny part of all the objects in space. Yet this is more than you could count in a lifetime. Space is filled with countless different groups of objects. So where in the vast region of space is Earth? Earth is part of one small group called the Solar System.

LAR SYSTEM?

TECHTREK
myNGconnect.com

Student eEdition
Vocabulary Games
Digital Library
Enrichment Activities

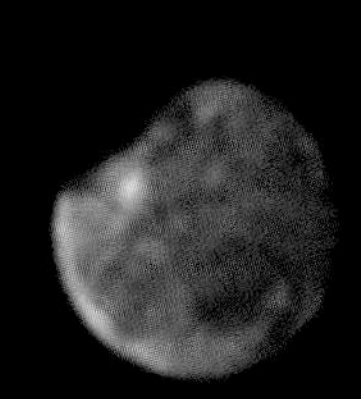

Jupiter and its largest moon Ganymede are part of the Solar System.

SCIENCE VOCABULARY

star (STAR)

A **star** is a ball of hot gases that gives off light and other types of energy. (p. 110)

The twinkling objects in the night sky are great balls of gas and dust called stars.

universe (YŪ-ni-vurs)

The **universe** is everything that exists throughout space. (p. 111)

The universe is so large that scientists have not yet determined its exact size.

galaxy (GA-luk-sē)

A **galaxy** is a star system that contains large groups of stars. (p. 112)

This starburst galaxy, M82, is 12 million light-years from Earth.

dwarf planet (dworf PLA-nit)

galaxy (GA-luk-sē)

moon (MÜN)

planet (PLA-nit)

star (STAR)

universe (YŪ-ni-vurs)

planet (PLA-nit)

A **planet** is a large nearly round space object that orbits a star. (p.116)

Venus is one of the eight planets in the solar system.

moon (MÜN)

A **moon** is a large rocky object that orbits a planet. (p. 120)

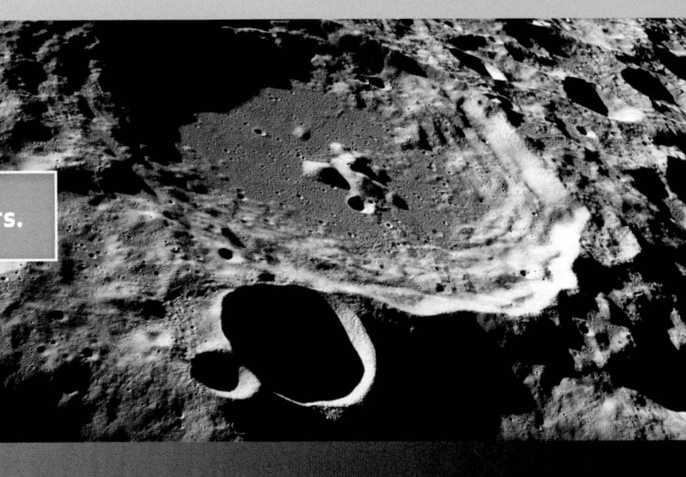

Earth's moon is covered with craters.

dwarf planet (dworf PLA-nit)

A **dwarf planet** is an object that orbits the sun, is larger than an asteroid and smaller than a planet, and has a nearly round shape. (p. 129)

A dwarf planet is neither a planet nor a satellite.

Stars and Galaxies

Stars What do you see in this clear night sky? Except for the distant glow from city lights, all the light in this night sky comes from stars. A **star** is a ball of hot gases that gives off light and other types of energy. New stars are born as old ones die. An average star such as our sun will live for billions of years.

The sun is the nearest star to Earth. Like all stars, the sun formed from a huge cloud of dust and gas called a nebula. When stars form, they can grow to many sizes. The smallest ones are the size of planets. The sun is an average-sized star. Still, it's huge. If the sun were a hollow ball, more than a million Earths could fit inside it! The largest stars are many times bigger than the sun. But except for the sun, stars are so far from Earth, they look like points of light.

This view of the night sky is seen from Arizona.

Stars look close together in the sky. But actually, they are separated by vast distances in space. It's like comparing the distance between the stop sign and the high-wire towers in the picture. The towers look like they are just a little to the left and a little to the right of the stop sign. But you know it would be quite a hike to walk from the sign to one of the towers. Also the distance between the towers is greater than it looks. All three objects look closer to each other than they really are because of your viewpoint.

It's the same thing with stars. They look close together because of our viewpoint from Earth. However, unlike the sign and towers, stars aren't just a kilometer or two apart. Most stars are trillions of kilometers away from each other! Between the stars is some gas and dust but mostly empty space. All of the stars, gas, dust, other objects, and empty space make up the **universe**. The universe contains everything that exists.

Galaxies Not all of the points of light you see in the night sky are single stars. Powerful telescopes show that some of these points of light are galaxies. A galaxy is a system of stars, gas, and dust held together in space by gravity. Gravity is a force that pulls any two objects toward each other. Many stars have objects moving around, or orbiting, them. Gravity holds these objects in orbit.

The universe contains billions of galaxies in different shapes and sizes. The three main shapes are spiral, elliptical, and irregular.

If you could look down on a spiral galaxy from above, it would look like a pinwheel. Spiral galaxies have a center that is packed with thousands of stars. Moving out from the center are arms that contain many more stars. The arms also contain most of the gas and dust where new stars form.

Stars are forming in the clouds of gas and dust in this irregular galaxy.

Find the arms of this spiral galaxy.

Earth is located on an arm of a spiral galaxy called the Milky Way. You can't see the spiral shape because Earth is part of the galaxy. But you can see part of the arm on a clear night. It looks like a white band or pathway across the sky. That is how the Milky Way got its name.

Elliptical galaxies are round or oval with no arms. Elliptical galaxies contain little gas or dust. These materials may have been used up long ago to form the stars in the galaxy.

All other galaxies are called irregular because they have no regular shape. Irregular galaxies contain many clouds of gas and dust in which stars are forming.

Messier 82 (M82) is the name of this elliptical galaxy. The red globs are clouds of hydrogen gas blasting out of the center.

Before You Move On

1. What is a star?
2. What type of galaxy is the Milky Way?
3. **Draw Conclusions** Which type of galaxy is least likely to have many stars forming? Why?

NATIONAL GEOGRAPHIC

CONSTELLATIONS

Have you ever connected dots on a page to make a picture? The same idea applies to constellations. A constellation is a group of stars that form a recognizable pattern or shape in the sky.

Constellations can be a useful tool. Without modern calendars, ancient peoples relied on star patterns to mark the changes of the seasons. Farmers would decide by constellations when to plant and harvest crops. People also used constellations to explain natural events or religious beliefs. Travelers by land and sea would use the familiar star patterns to find their way around. Even today, stars are used to navigate. In which direction are you looking if you can see Polaris, the north star? You are facing north, of course!

This constellation is called Leo. Leo is Latin for lion. It can be easily seen in spring in the Northern Hemisphere.

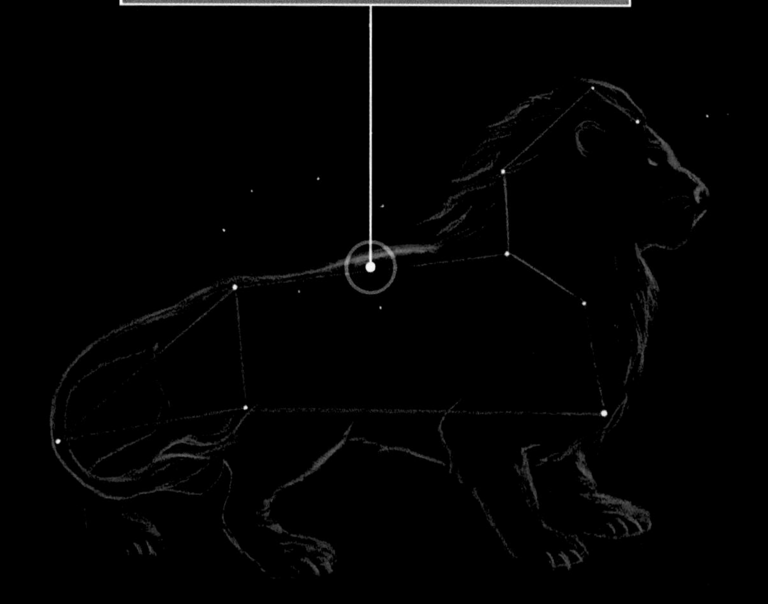

Not every ancient civilization saw the same shape in the sky. The ancient Romans saw a giant man in the grouping of stars known as Orion. Some Native Americans, such as the Lakota, saw a giant hand in the stars that make up Orion's belt. The Dogon people of Africa saw the stars of Orion's belt as a staircase.

Whatever shape is seen, it only looks this way if you are standing on Earth. The constellation appears flat, like words on a page. But if you looked at the stars that make up Orion from space, the shape would be completely different.

Constellations are a human invention. Today, 88 different constellations are recognized by astronomers. The constellations seen in the Northern Hemisphere of Earth are different from those seen in the Southern Hemisphere. In the Northern Hemisphere, most are based on ancient Roman and Greek mythology. In the Southern Hemisphere the constellations are more modern. They are generally familiar objects or animals, such as a compass or a porpoise.

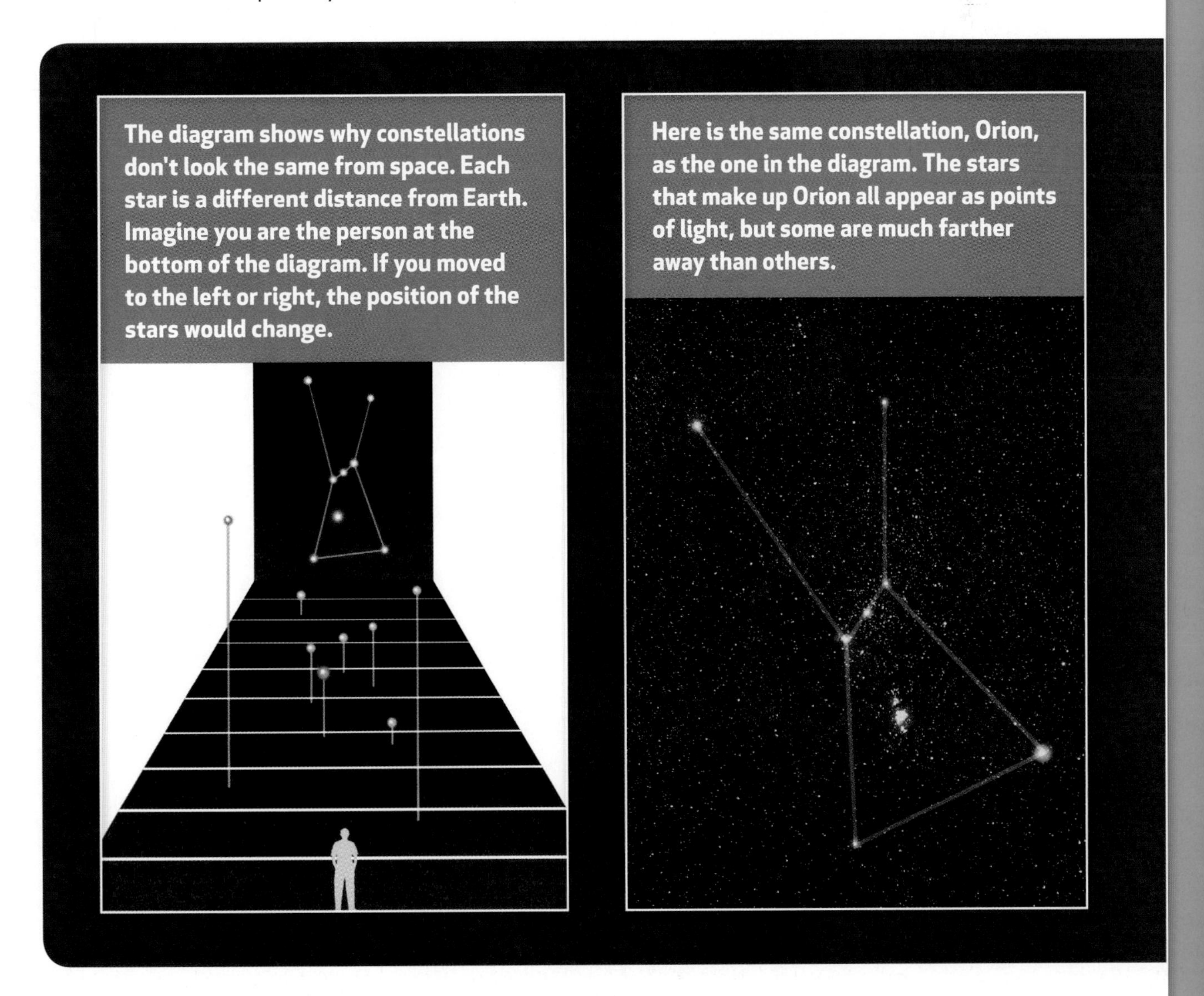

The diagram shows why constellations don't look the same from space. Each star is a different distance from Earth. Imagine you are the person at the bottom of the diagram. If you moved to the left or right, the position of the stars would change.

Here is the same constellation, Orion, as the one in the diagram. The stars that make up Orion all appear as points of light, but some are much farther away than others.

The Solar System

A galaxy is a star system that includes millions of stars. Some star systems, however, have only one star. Within a galaxy, many stars are each at the center of their own star system. One of these star systems is our own solar system. It includes our closest star—the sun—and all the space objects that revolve around it.

Planets are the largest space objects that orbit a star. Our solar system contains eight planets. The diagram shows the order of the planets from the sun. But the distances between the sun and each of the planets are much greater than shown here. Each planet is shaped generally like a ball, but it is not completely spherical, or round. For example, Earth is slightly wider from west to east than it is from north to south. It's like a slightly flattened ball.

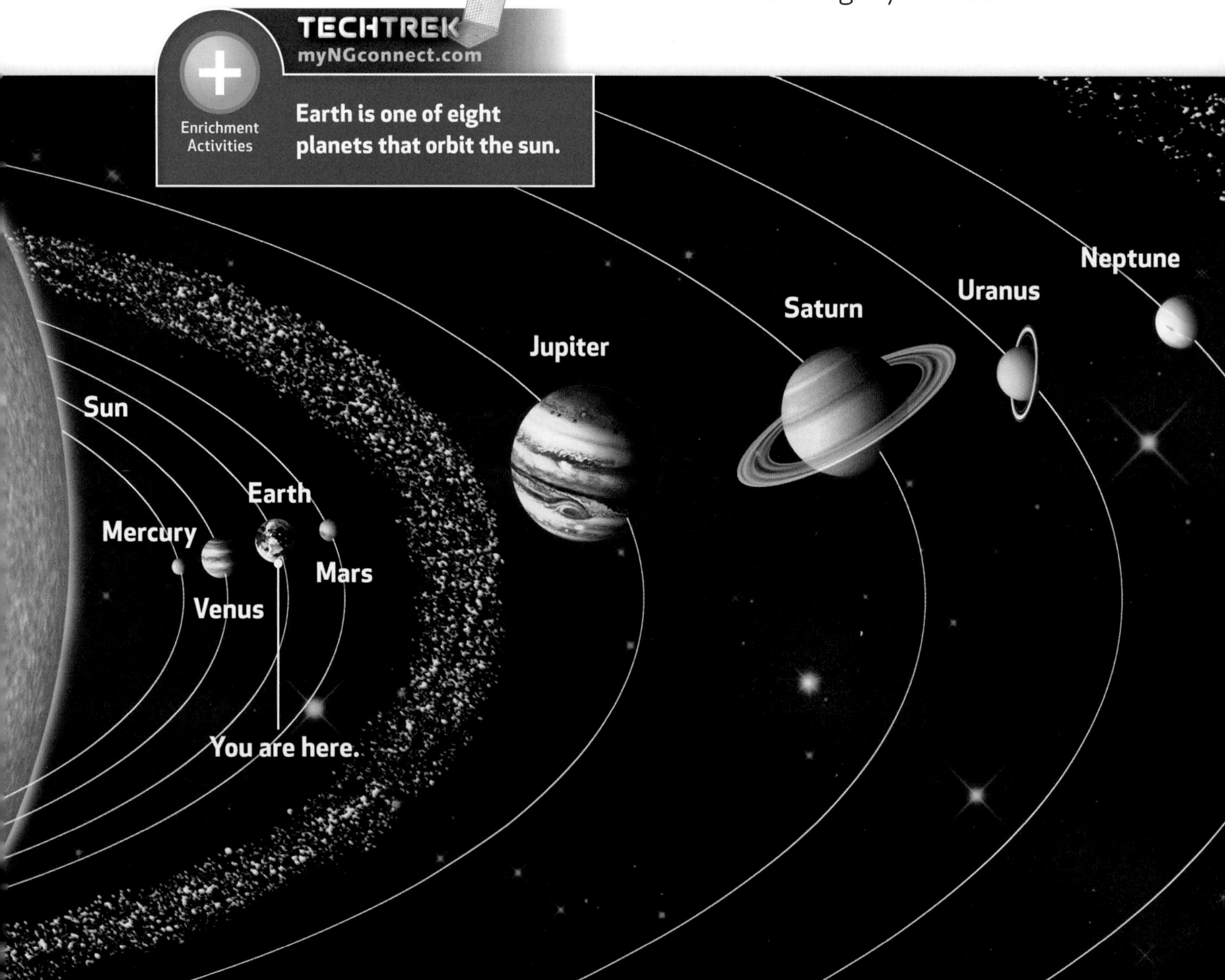

The eight planets are separated into two groups—the inner planets and the outer planets. The inner planets are closest to the sun. They are made mostly of rock, and are also called the rocky planets. Earth is an inner planet.

The outer planets are farthest from the sun. They are made mostly of gases surrounding a smaller rocky core. The outer planets are the largest in the solar system and are also called the gas giants.

Science in a Snap! How Far Apart

Mercury	0.4
Venus	0.7
Earth	1.0
Mars	1.5
Jupiter	5.0
Saturn	9.5
Uranus	19.0
Neptune	30.0

Consider that Earth is 1.0 unit from the sun. The chart shows how many of the same units each of the other planets are from the sun. Choose an object to represent one unit. Use your object to make a scale model.

Make markers labeled Sun and with each planet's name. Use the unit you chose to place the markers on a line.

What do you notice about how far the planets are from one another?

Before You Move On

1. Define what a planet is.
2. How are the inner planets different from the outer planets?
3. **Infer** A year on a planet is how long it takes the planet to orbit the sun once. Why do the planets have years of different lengths?

The Inner Planets

Mercury Mercury is a planet of extremes. It's the smallest, fastest, and closest planet to the sun. It has the oldest surface and the wildest temperature changes.

What would it be like on Mercury's surface? The sun would look much bigger than it does from Earth. The landscape would be covered with deep craters caused by space rocks, such as asteroids, crashing into the planet.

Even with a spacesuit, you would not be able to leave your spacecraft. Mercury has no atmosphere to protect it from the sun's radiation. Also the temperature during the day is extremely hot, and the nights are extremely cold.

Venus Venus is the second planet from the sun and Earth's closest neighbor. It is almost the same size as Earth, too. When you look into the night sky, only the moon is brighter than Venus.

Venus is probably the last place you would want to land a spacecraft. The atmosphere is made mostly of carbon dioxide. It is extremely thick and poisonous. Lightning, high winds, and difficulty seeing would make landing hard.

Once there, you could not step outside. The pressure of the atmosphere is enormous—strong enough to crush a car! Venus' thick atmosphere also makes it the hottest planet—hot enough to melt lead. This could explain why water doesn't exist on the surface of Venus. Any water would have boiled away.

On Venus the Sun appears to rise toward the west and appears to set toward the east. Can you guess why? Venus rotates in the opposite direction to Earth.

Venus is blanketed in dense clouds of sulfuric acid. This image of Venus's surface was put together using radar waves. The radar waves let astronomers see through the dense clouds.

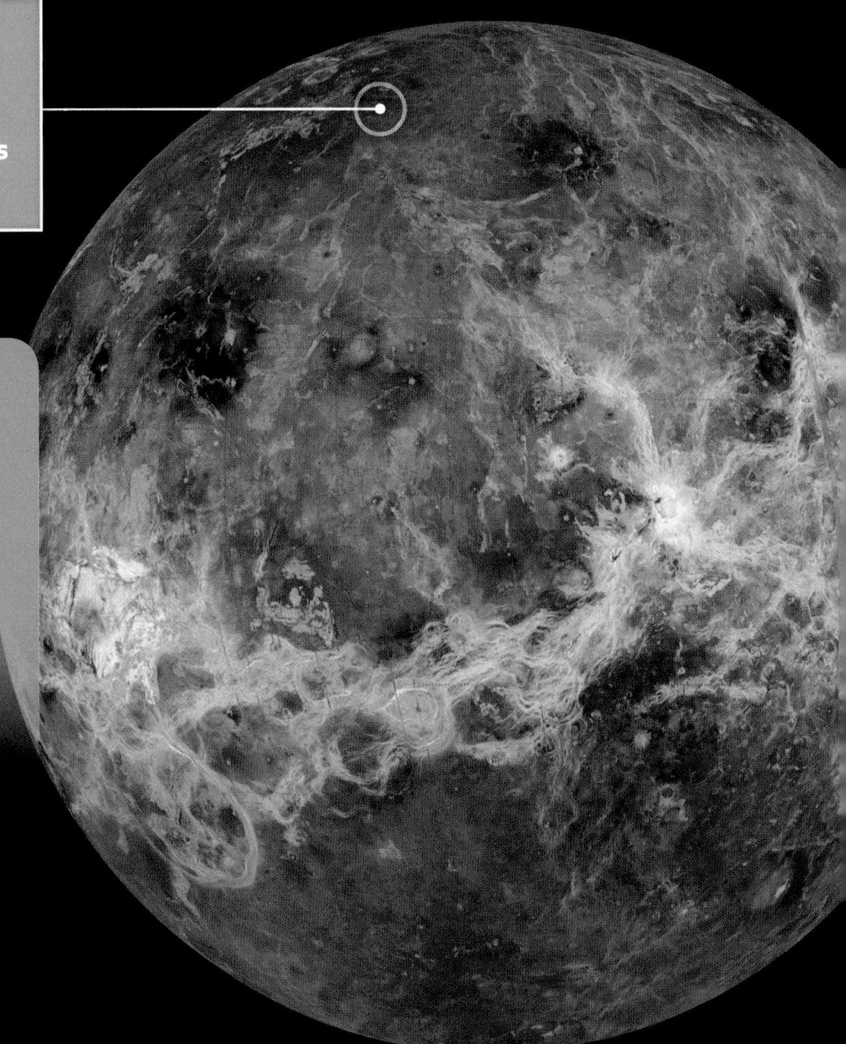

VENUS FACTS

Number of Moons: 0

Length of Rotation: 243 Earth days

Length of Revolution: 225 Earth days

Surface Temperature:
Average: about 465°C (870°F)

Earth and Its Moon

Earth is the third planet from the sun. It is the home of all known life in the universe. Two basic features make life on Earth possible. The first is Earth's distance from the sun. If Earth were closer to the sun, it would be too hot. If Earth were farther away, it would be too cold. The second feature is the presence of liquid water. All living things need water. No other planet has these two features.

Earth's atmosphere is also important to life. The atmosphere is a mixture of gases that surrounds the planet. It extends more than 600 kilometers (about 372 miles) into space. Most of the gases in Earth's atmosphere support, protect, or do no harm to living things.

Earth has one **moon**. A moon is a large rocky object that orbits a planet. Earth's moon has almost no atmosphere. As a result, just about the only time the moon's surface changes is when it gets struck by a rock from space.

With no atmosphere to protect it from space rocks, the moon's surface is covered with craters.

Earth is sometimes called the "blue planet" because most of its surface is covered with water.

EARTH FACTS

Number of Moons: 1

Length of Rotation: 24 hours or 1 day

Length of Revolution: 365 days

Surface Temperature:
Hottest: 58°C (136°F)
Coldest: -88°C (-126°F)

Mars and Its Moons

The fourth planet from the sun is Mars. It's only half the size of Earth, but in many ways Mars is more like Earth than any other planet in the solar system. Mars and Earth both have polar ice caps and four seasons. If you landed on Mars' surface, you would see land features similar to some of the places on Earth's surface. There are canyons, hills, plains, valleys, and volcanoes. In fact, Mars has the largest known volcano in the solar system, Olympus Mons.

Since the air on Mars is mostly carbon dioxide, you would need a spacesuit to go exploring. The atmosphere is thinner than Earth's. As a result it is always very cold, especially at night. While you walked around you would see that it is rocky and dry. Your spacesuit would become very dusty.

In the sky above Mars are Phobos and Deimos, its two tiny moons. Each is only about as wide as a large city. Some scientists think these moons may be asteroids that were caught by the gravity of Mars and pulled into orbit around it.

If red is your favorite color, then you might like it on Mars. Iron minerals in the soil give the land a rusty-red color.

MARS FACTS

Number of Moons: 2

Length of Rotation: about 1 Earth day

Length of Revolution: 687 Earth days

Surface Temperature:
Hottest: about –30°C (–22°F)
Coldest: about –100°C (–148°F)

There is growing evidence that Mars' surface once held liquid water.

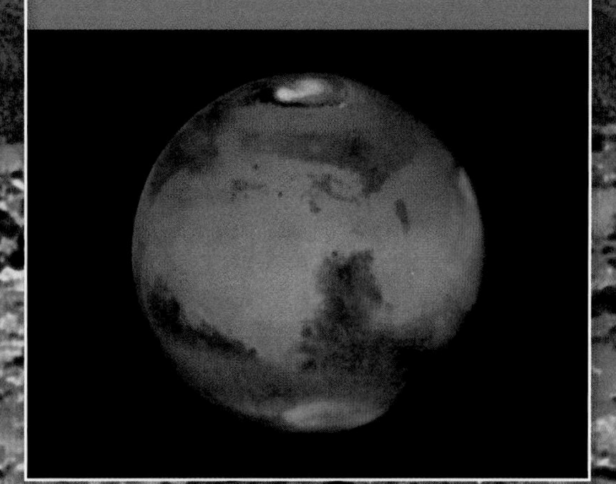

Before You Move On

1. What features of Earth make life possible?
2. Compare and contrast the atmospheres of Venus and Mars.
3. **Infer** Why do you think Mercury takes the least amount of time to travel around the sun?

The Outer Planets

Jupiter and Its Moons The fifth and biggest planet in the solar system is Jupiter. It's the first of the "gas giants." Jupiter is so large that 1,324 Earths could fit inside!

What would it be like if you went to Jupiter? For starters you would have nowhere to land. Jupiter is made of gas. There are no solid surfaces anywhere. Scientists do think Jupiter's core contains some rocky materials. The temperature of the core may be hotter than the surface of the sun.

Even though Jupiter is made of gas, you could not fly a spacecraft into it. The pressure is so great that any solid matter gets crushed.

Jupiter's atmosphere is mostly a mixture of hydrogen and helium gases. Its famous Great Red Spot is a rotating storm, much like a hurricane. It was first discovered more than 300 hundred years ago. Jupiter has a few rings too, but not like the other gas giants. Jupiter's rings are hard to see. They are thought to be the shattered pieces of two moons that smashed into each other.

Jupiter has a huge family of moons. The four largest are about the size of Mercury. Scientists often say that Jupiter is like a solar system all by itself. Each of the moons is different. It's fourth largest, Europa, has a thin shell of ice. But Io, Jupiter's third largest, is covered with active volcanoes and lakes of hot lava.

Jupiter's colorful streaks and bands are super-cooled clouds of water and ammonia.

JUPITER FACTS

Number of Moons: at least 62

Length of Rotation: 9 hours 56 minutes

Length of Revolution: 4333 Earth days (about 12 Earth years)

Surface Temperature:
Average: about −145°C (−229°F)

Saturn and Its Moons The sixth planet, Saturn, is almost twice as far from the sun as Jupiter. It is the second largest planet in the solar system.

As with all of the gas giants, you would have nowhere to land a spacecraft on Saturn—and you wouldn't want to. Like Jupiter, the pressure in Saturn's clouds would crush anything. The wind would tear a spacecraft apart. Wind speeds in Saturn can reach 500 meters (1600 feet) per second. That's more than three times faster than the most powerful tornadoes.

Saturn's rings make this planet easy to identify. The rings are made of billions of pieces of ice and rock. Scientists think they are the remains of moons or other objects.

Like Jupiter, Saturn has dozens of moons of many shapes and sizes. Some are located inside its rings. The largest moon, Titan, has an atmosphere.

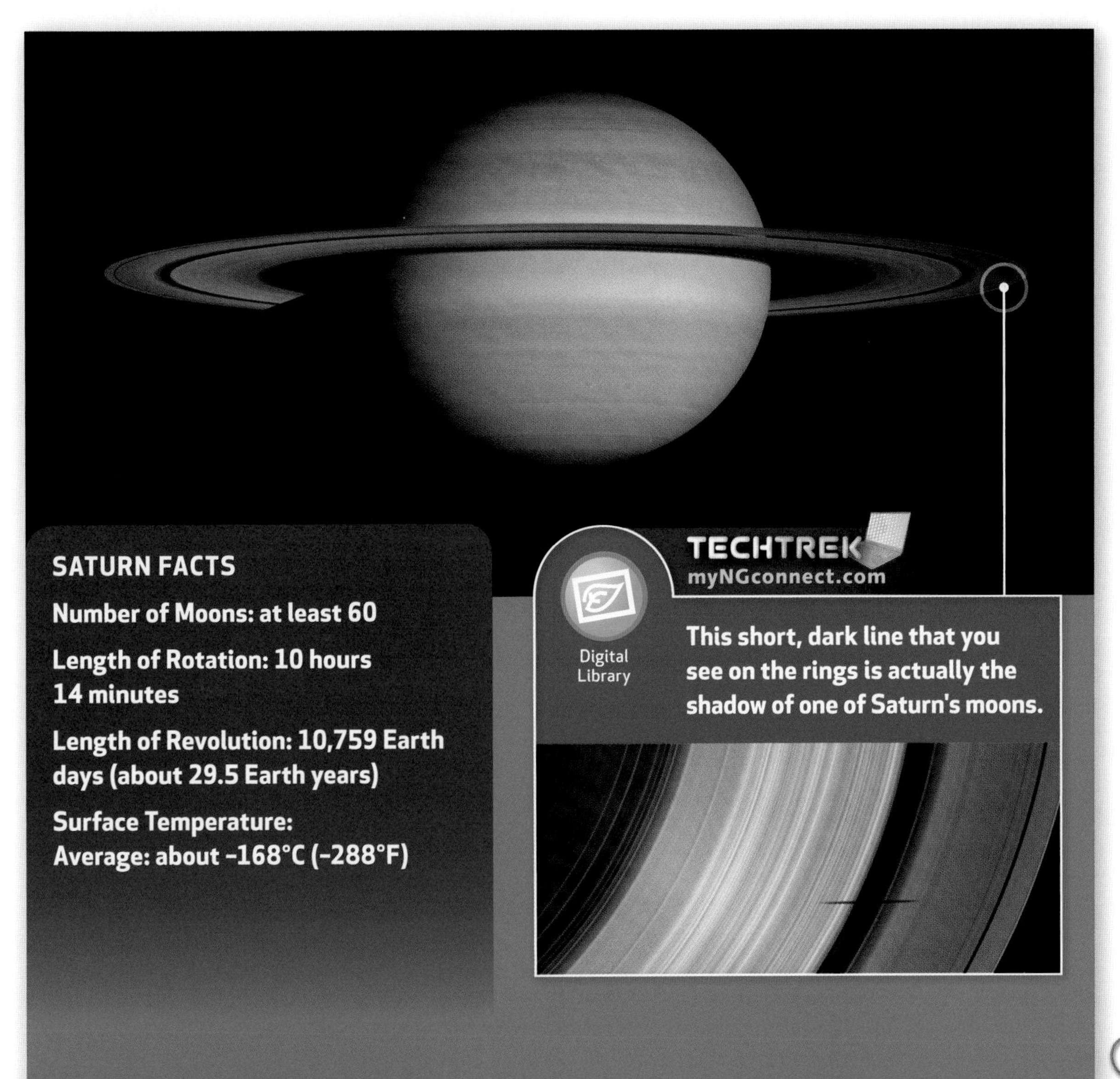

This short, dark line that you see on the rings is actually the shadow of one of Saturn's moons.

SATURN FACTS

Number of Moons: at least 60

Length of Rotation: 10 hours 14 minutes

Length of Revolution: 10,759 Earth days (about 29.5 Earth years)

Surface Temperature: Average: about -168°C (-288°F)

Uranus and Its Moons The seventh planet from the sun is Uranus. More than four times larger than Earth, Uranus is the third-largest planet in the solar system.

Uranus' atmosphere is mostly hydrogen and helium gases. An outer layer of frozen methane gas surrounds the planet. This gives it its blue-green color. Scientists believe there may be an ocean of liquid water deep within Uranus's atmosphere. But it is unlikely this ocean could support life. Below the ocean is a rocky core about the same size as Earth.

Uranus is tilted on its side. So instead of spinning like a top, it appears to roll like a ball. One idea why is that Uranus was hit by a large space object. The force of the impact tilted the planet.

For decades scientists thought Uranus had only four moons. But when the Voyager 2 spacecraft passed by Uranus in 1986, scientists were excited to find 10 more! Since then, powerful telescopes have helped scientists find even more moons around Uranus. They are difficult to spot from nearly 3 billion km (1.86 million miles) away.

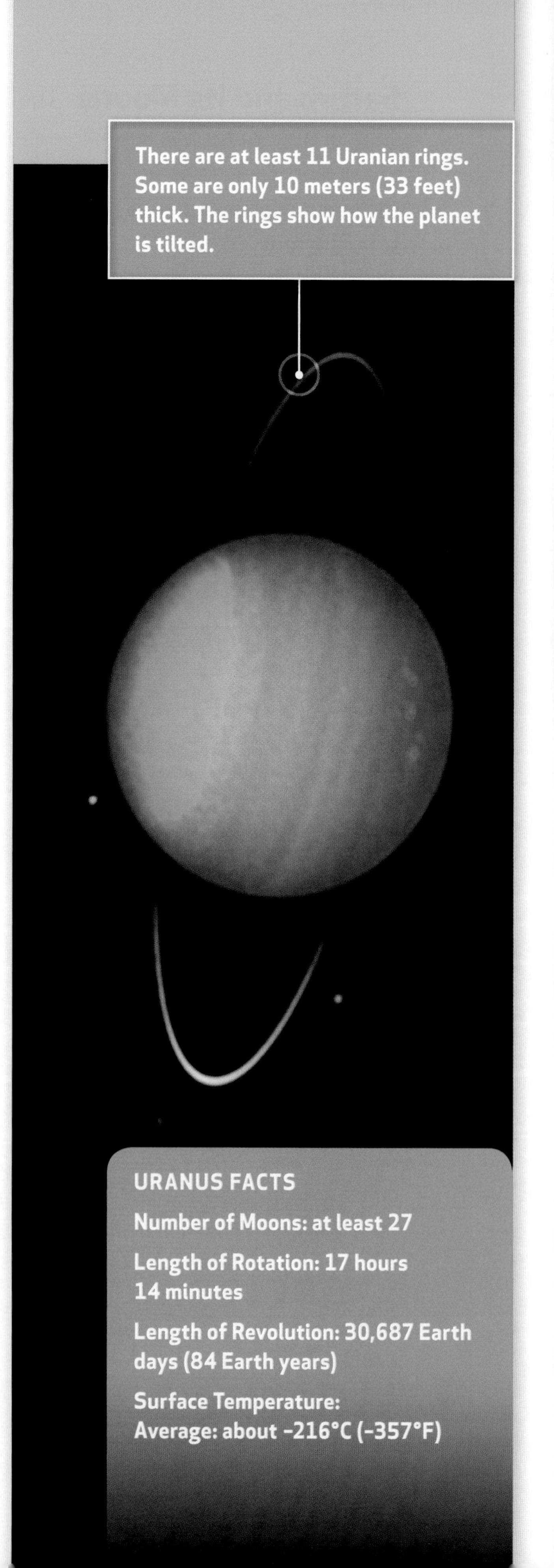

There are at least 11 Uranian rings. Some are only 10 meters (33 feet) thick. The rings show how the planet is tilted.

URANUS FACTS

Number of Moons: at least 27

Length of Rotation: 17 hours 14 minutes

Length of Revolution: 30,687 Earth days (84 Earth years)

Surface Temperature:
Average: about -216°C (-357°F)

Neptune and Its Moons

The eighth and last planet in the solar system is Neptune. It is so similar to Uranus that some scientists call them twins. Neptune, like Uranus, has a cold outer layer of methane gas. This gives the planet its blue color. Scientists think there may be an ocean of liquid water in Neptune, too. The water is super-heated, but the pressure in Neptune's atmosphere keeps the water from boiling away.

All of the gas giants are windy, but Neptune's winds are the fastest in the solar system. Winds can blow more than 550 meters (1800 feet) per second.

Neptune is known to have at least 13 moons. The largest is Triton. It orbits in a direction opposite Neptune's rotation. It is the only moon in the solar system to do this. Scientists think that Triton formed outside of the solar system and was pulled in by Neptune's gravity.

Neptune has a few rings too, but they are very faint. The rings are made of particles of dust.

NEPTUNE FACTS

Planet Name: Neptune

Number of Moons: at least 13

Length of Rotation: 16 hours 7 minutes

Length of Revolution: 60,190 Earth days (165 Earth years)

Surface Temperature:
Average: about -214°C (-353°F)

Before You Move On

1. Describe the atmosphere of Jupiter.
2. Explain how Uranus may have come to be tilted on its side.
3. **Generalize** What are at least four features the gas giants all have in common?

Other Objects in the Solar System

The sun, planets, and moons are not the only objects in the solar system. Other objects include asteroids, dwarf planets, and comets.

Asteroids Asteroids are rocky objects that orbit the sun. They are smaller than dwarf planets. Asteroids range in size from 1,000 km (600 miles) across to as small as a grain of sand. About 50,000 asteroids have been studied so far.

Most asteroids are in the space between Mars and Jupiter. Because it is filled with tens of thousands of asteroids, this area is called the asteroid belt. Scientists have a few ideas about how the asteroid belt formed. Some suggest the asteroids are pieces of a lost planet.

Millions of asteroids travel in a beltway between Mars and Jupiter in this illustration.

About 200 asteroids in the asteroid belt are more than 97 kilometers (60 miles) wide. That's wider than the state of Rhode Island.

Others suggest the asteroids formed when several large objects crashed into each other. A third idea says that they are simply left-over material from the formation of the solar system. So far, none of these ideas has been proven or rejected.

Some asteroids travel outside of the main asteroid belt. Their orbits carry them into the area between the orbits of Mars and Earth.

Scientists think some of these asteroids broke away from the asteroid belt. One of the largest of these is Eros. In 2012, this asteroid is expected to pass within 26.7 million km (about 16.6 million miles) of Earth. This is considered to be a very close pass.

The asteroid, Eros is about 34 kilometers long. In 2001 a spacecraft orbited the asteroid and then landed on it. The spacecraft sent back information to Earth.

Dwarf Planets From 1930 until 2006, the solar system had a ninth planet called Pluto. Then, in 2006, an international meeting of scientists changed the definition of a planet. Pluto had to be reclassified.

Recall that a planet is the largest type of nearly round space object to orbit a star. Other than the sun itself, the eight planets are the largest objects in the solar system. But Pluto is much smaller than Mercury, and smaller than our moon. Most scientists agree that Pluto is not large enough to be a planet.

Since Pluto's discovery in 1930, it has travelled only one-third of its way around the sun.

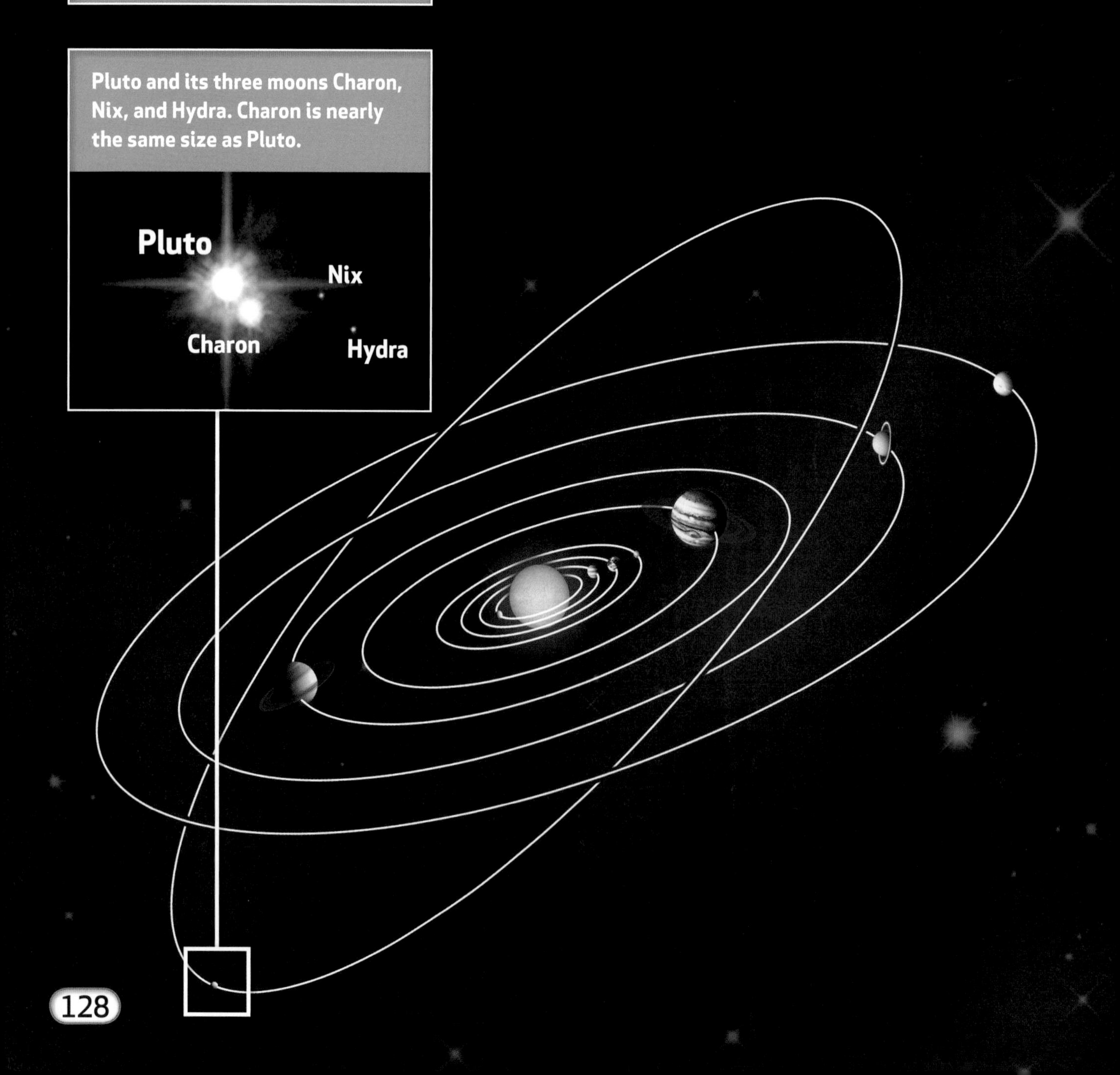

Pluto and its three moons Charon, Nix, and Hydra. Charon is nearly the same size as Pluto.

Pluto's size makes it a **dwarf planet** A dwarf planet is an object that orbits the sun, is larger than an asteroid and smaller than a planet, and has a nearly round shape. Our solar system has five known dwarf planets.

Pluto is mainly ice. It has a thin atmosphere, mostly of methane gas. Pluto is also extremely cold. If you could stand on Pluto's surface, the sun would look dim and small because it is so far away.

Compared with the planets, Pluto's orbit is unusual. It moves at an angle to the eight planets. At certain times Pluto crosses inside Neptune's orbit.

One dwarf planet, Ceres, used to be considered a large asteroid. It travels in the asteroid belt and shares its orbit with thousands of asteroids. But in 2006, scientists decided it was actually a dwarf planet. Ceres is about 950 km (590 miles) wide—wider than Nevada. Ceres' size means it is large enough to have strong gravity. Its stronger gravity holds it in the shape of a sphere. For these reasons, Ceres was classified as a dwarf planet.

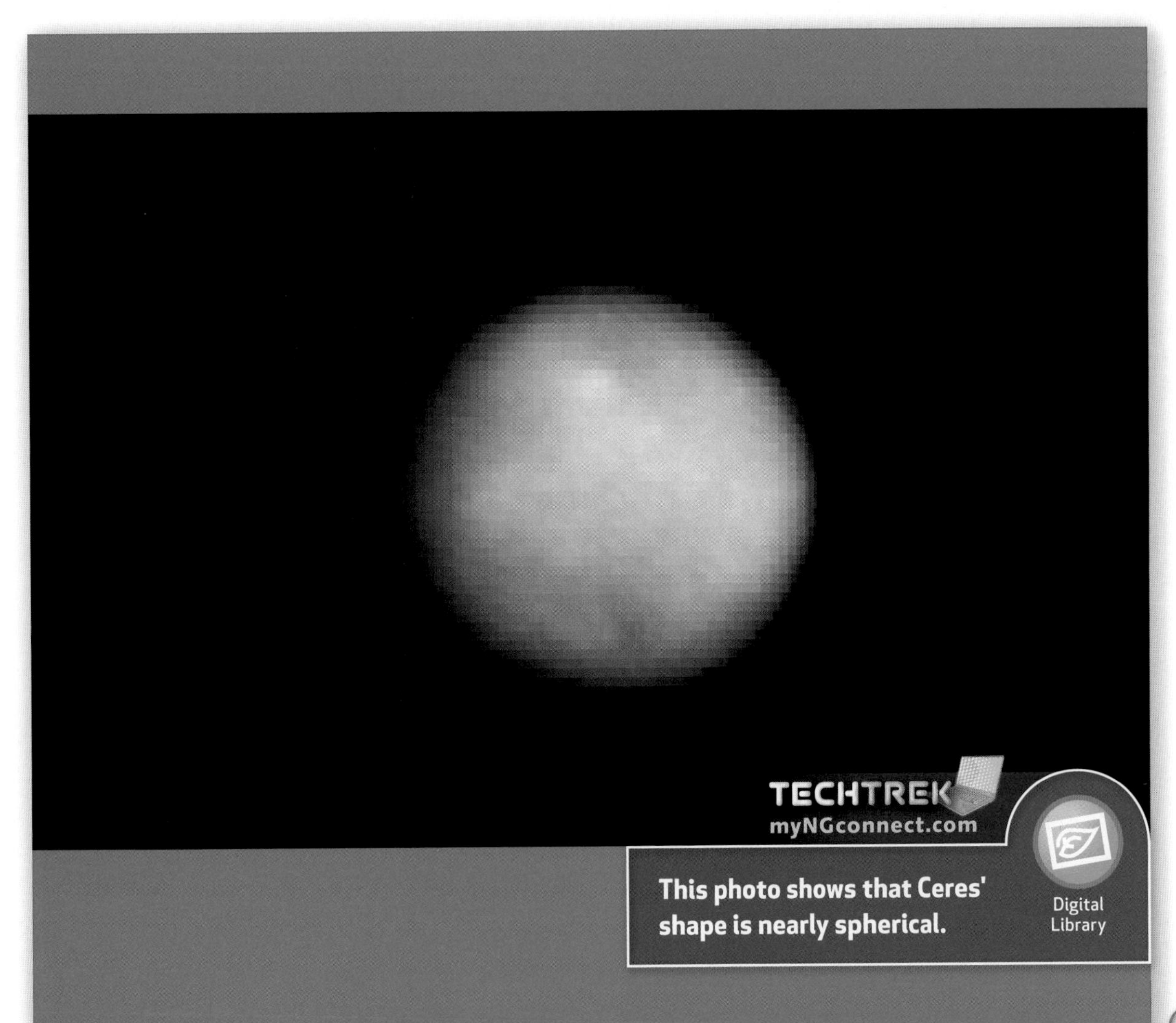

This photo shows that Ceres' shape is nearly spherical.

Comets One of the more unusual objects in space is a comet. Imagine a ball of ice and rock flying through space. That is basically a comet. When comets pass close to the sun, the sun's energy changes some of the ice to gas. The comet begins to glow. The glowing part is called a coma. A trail of gas and dust streams out behind it. The stream is called a tail. The tail can extend for millions of miles into space! Eventually, the comet travels far enough from the sun that it refreezes. The comet stays frozen until the next time it passes close to the sun.

Comets leave behind a trail of rocky dust. Planets often pass through these trails. On Earth, passing through the trail left by a comet makes one of the most amazing sights in the night sky. These are meteor showers, also known as shooting stars. Meteor showers are thousands of bits of solid material hitting Earth's atmosphere and burning up. Every year, Earth passes through the remains of at least 12 comets.

Comets orbit the sun in an area beyond Neptune called the Kuiper (KĪ•pur) belt. Some comets have short orbits. They return close to Earth often enough to be recorded in history. Halley's Comet returns every 76 years.

As the glowing comet continues its orbit around the sun, the tail points away from the sun.

Pieces of asteroids or comets that fall through space and enter Earth's atmosphere are called *meteoroids*. Meteoroids are smaller than comets and asteroids. Friction between a meteoroid and Earth's atmosphere causes heat first, then light. The light we see streaking across the sky is a *meteor*, or shooting star.

If an asteroid, comet, or meteoroid does not burn up completely on its trip through the atmosphere, it can land on Earth's surface. These pieces of space rock are called *meteorites*.

This meteor shower was caused by the comet Tempel-Tuttle.

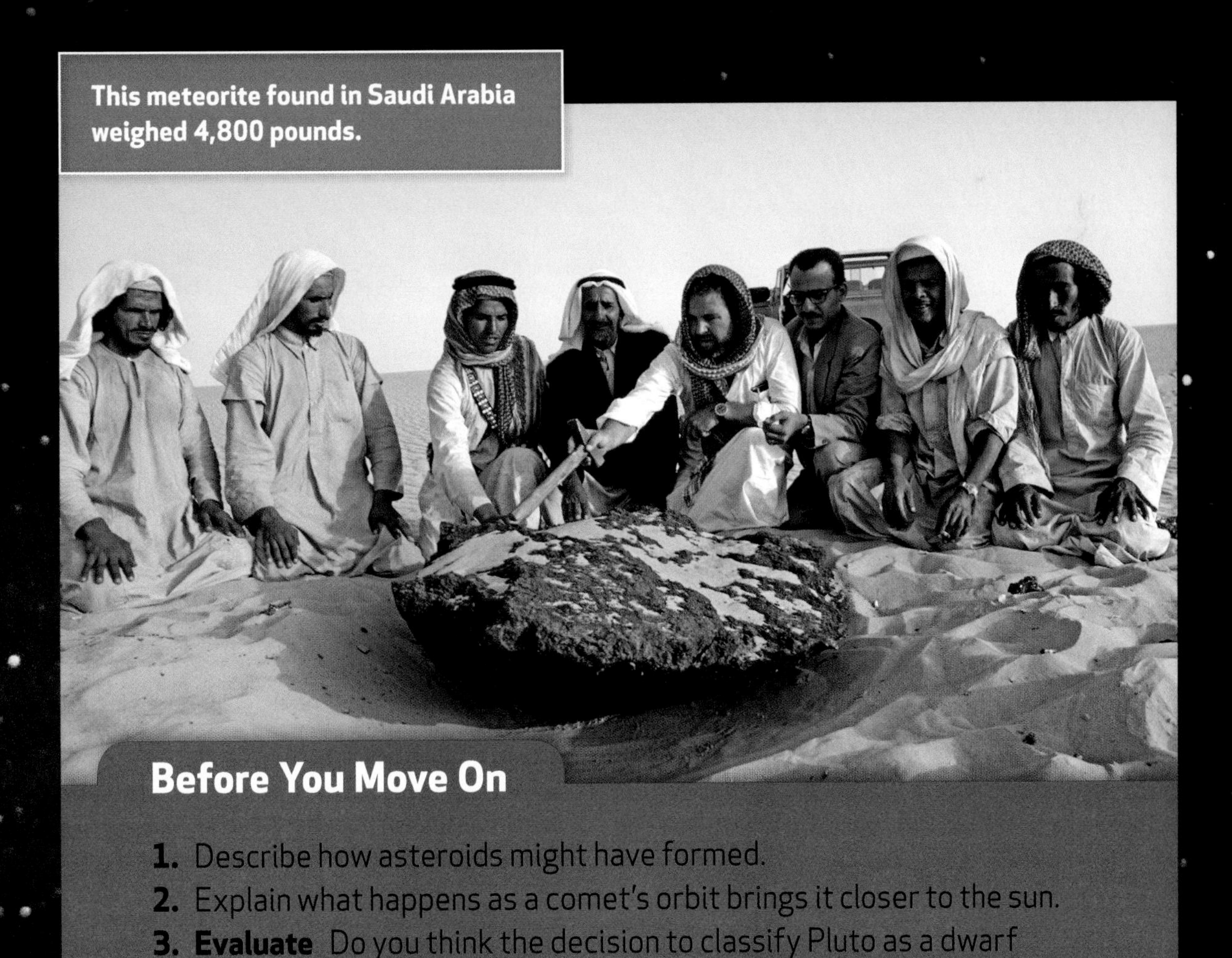
This meteorite found in Saudi Arabia weighed 4,800 pounds.

Before You Move On

1. Describe how asteroids might have formed.
2. Explain what happens as a comet's orbit brings it closer to the sun.
3. **Evaluate** Do you think the decision to classify Pluto as a dwarf planet was correct or incorrect? Explain your reasoning.

Conclusion

The solar system is within the Milky Way Galaxy. It is one of countless star systems throughout the universe. The sun and everything else in the solar system formed within a nebula billions of years ago. This includes the eight planets, a few dwarf planets, and many moons, asteroids, and comets.

Big Idea Our solar system is made up of everything that revolves around the sun, including planets, dwarf planets, moons, asteroids, and comets.

THE SUN

PLANETS AND MOONS

DWARF PLANETS, ASTEROIDS, AND COMETS

Vocabulary Review

Match the following terms with the correct definition.

A. galaxy
B. planet
C. moon
D. universe
E. star
F. dwarf planet

1. Everything that exists throughout space
2. A star system that contains large groups of stars
3. A gaseous sphere that gives off light and other types of energy
4. A large, round, rocky object that orbits a planet
5. An object that orbits the sun, is larger than an asteroid and smaller than a planet, and has a nearly round shape
6. A large nearly round space object that orbits a star

Big Idea Review

1. **List** List the eight planets in order from the sun.

2. **Describe** Identify and describe one gas giant other than Jupiter.

3. **Compare and contrast** Explain how the inner and outer planets are alike and different.

4. **Interpret** Why is the sun considered to be an average star?

5. **Infer** We can't see the spiral shape of our galaxy. How is this like not being able to see the shape of a forest you are in?

6. **Draw Conclusions** Some scientists like to say that Jupiter is almost a solar system all by itself. Why isn't Jupiter a solar system?

Write About the Solar System

Explain This is a photo of Mercury's surface. What features can you identify? Why does the surface look like this?

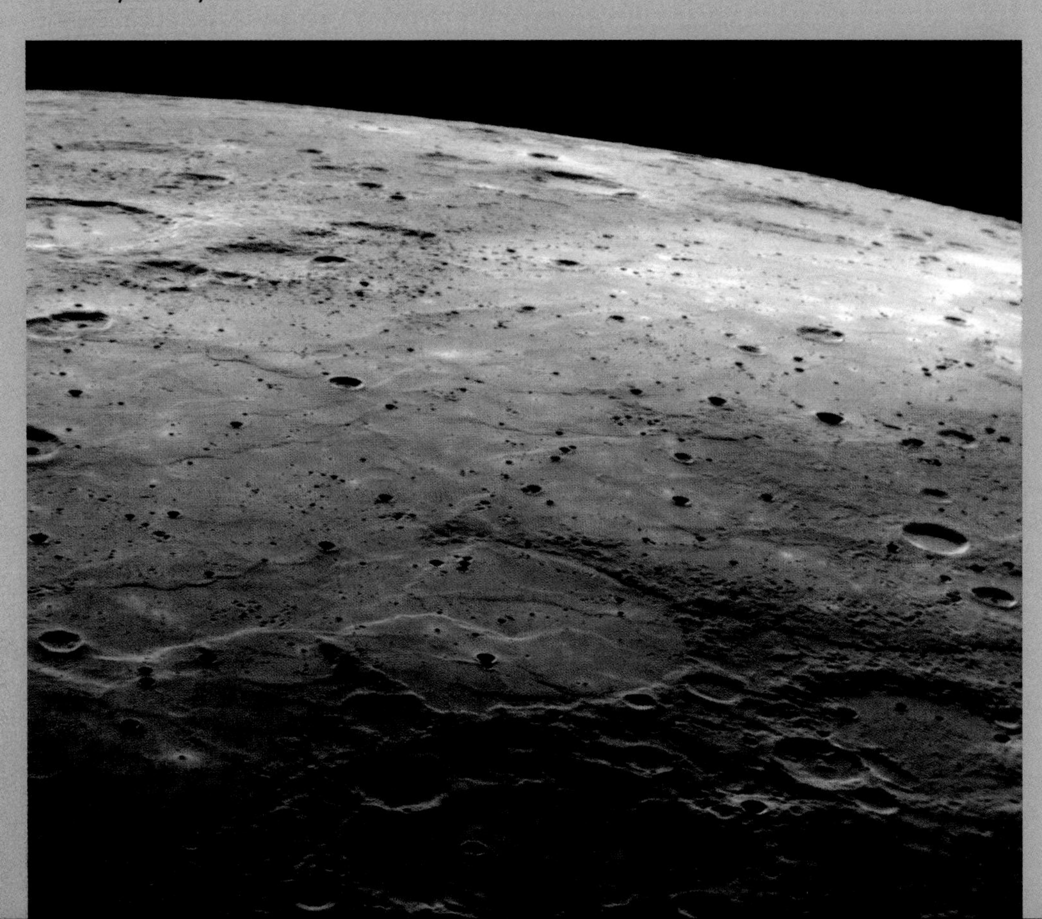

Do you want to know about space rocks? Ask a geologist who studies meteorites.

Have you ever wished on a shooting star? Do you wonder if they ever land on Earth? We might not all get to space someday, but many of us could have the opportunity to study objects that come from beyond the solar system. Dr. Meenakshi Wadhwa from Arizona State University is just the person to ask about space rocks. She is in charge of a huge collection of meteorites—over 1550 of them.

What do you do?

I teach students at a university and conduct research about rocks from space. I am also in charge of the Center for Meteorite Studies. Our center contains the largest meteorite collection at a university in the world.

Do you see a strong connection between what you do and Earth science?

There is a very strong connection! The basic ideas of Earth science form the foundation of my research. My scientific work involves studying rocks from other places in our solar system, such as meteorites and moon rocks. I also apply a lot of ideas from other physical sciences, such as physics and chemistry.

What would you say has been the coolest part of your job?

I would say that would be getting to interact with students who are really excited to learn and helping them make new discoveries. •

TECHTREK
myNGconnect.com
Student eEdition
Digital Library

Meenakshi Wadhwa looks at a thin slice of a meteorite through a microscope to see what the meteorite is made of.

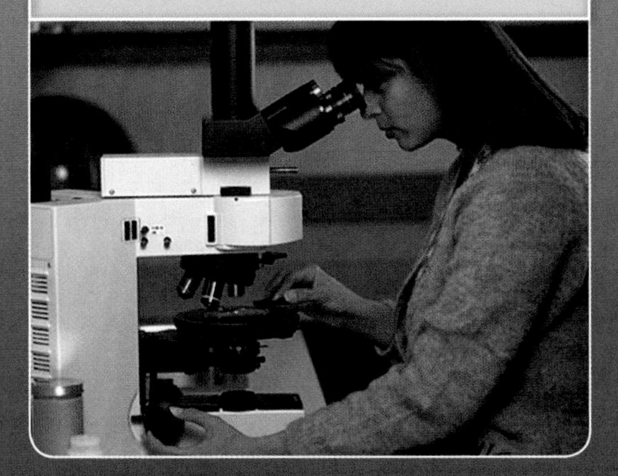

A smooth slice through a meteorite shows the different kinds of rocky material that make it up

What has been your greatest accomplishment so far?

One accomplishment has been in my scientific research. My work has helped advance our understanding about the timing of the formation of our solar system. Another accomplishment has been to motivate some of my students to go into careers in the sciences and in science education.

What did you like about science in elementary school?

When I was young, I had questions about everything, and science in school allowed me to understand the natural world around me. I didn't have a specific view of where I would end up. But I did know that I would be doing something fun, where I would get to apply my interest in the sciences.

What did you study in school and in college? Did you continue to study after college?

In school I studied a variety of subjects, including languages, geography, and the basic sciences. In college I majored in geology and had minors in physics and chemistry. After college, I went to graduate school and earned a doctorate (PhD) in Earth and Planetary Sciences.

Digital Library

Wadhwa examines one of the many large meteorites from the meteorite vault at Arizona State University.

NATIONAL GEOGRAPHIC

BECOME AN EXPERT

Asteroids and Comets: Friends or Foes?

Asteroids and comets have been crashing into **planets** ever since the sun and planets were born billions of years ago. All you need to do is look at Mercury or Mars to see the scars left by such impacts. The surface of Earth's **moon** is largely shaped by these impacts. What about Earth? Are we safe from these events? Not exactly.

About 5000 years ago in Australia, a giant asteroid or comet broke up over the sky. Its huge pieces fell like bombs, exploding into Earth's surface. What is left today is a cluster of 13 craters. Events like this one have happened many times in Earth's long history.

The famous Barringer Meteor Crater in Arizona is 1,219 meters (4,000 feet) wide and 168 meters (550 feet) deep.

planet

A **planet** is a large nearly round space object that orbits a star.

moon

A **moon** is a large object that orbits a planet.

Craters are dents in the ground caused by the force of a meteor's or comet's impact. More than 200 craters have been found on Earth so far. Why so few compared to the thousands of craters on the moon? Unlike the moon, Earth has an atmosphere. Most space rocks that would strike Earth burn up as they rub against the gas particles in the atmosphere. Some asteroids or comets have been large enough to make it through the atmosphere and form craters. But most craters have been worn away by weathering and erosion.

This crater was formed when a an asteroid or comet smashed into Australia thousands of years ago.

Rocks In Space Billions of asteroids and comets speed through the Milky Way **Galaxy**. Most comets orbit the sun near the outer edge of the solar system. Most asteroids orbit the sun between Mars and Jupiter. Scientists once thought Ceres was the largest asteroid. They now think it is big enough to be a **dwarf planet**.

Even though there are millions of asteroids in the asteroid belt, they are spread apart. If they were gathered together to form one large asteroid, it would still be smaller than the moon.

This illustration shows Ceres' nearly round shape. Some astronomers think there could be frozen water below its rocky surface.

galaxy
A **galaxy** is a star system that contains large groups of stars.

dwarf planet
A **dwarf planet** is an object that orbits the sun, is larger than an asteroid and smaller than a planet, and has a nearly round shape.

In 2001, a spacecraft called NEAR (Near Earth Asteroid Rendezvous) landed on one of the largest known asteroids, Eros. This was the first time a spacecraft orbited and landed on an asteroid. NEAR was able to take and send back to Earth 69 pictures of Eros before the spacecraft stopped working. The photos show that Eros has long grooves and many craters.

TECHTREK
myNGconnect.com
Digital Library

This painting shows NEAR approaching Eros for a landing. The blue squares are solar panels. They captured the sun's energy to operate the equipment onboard.

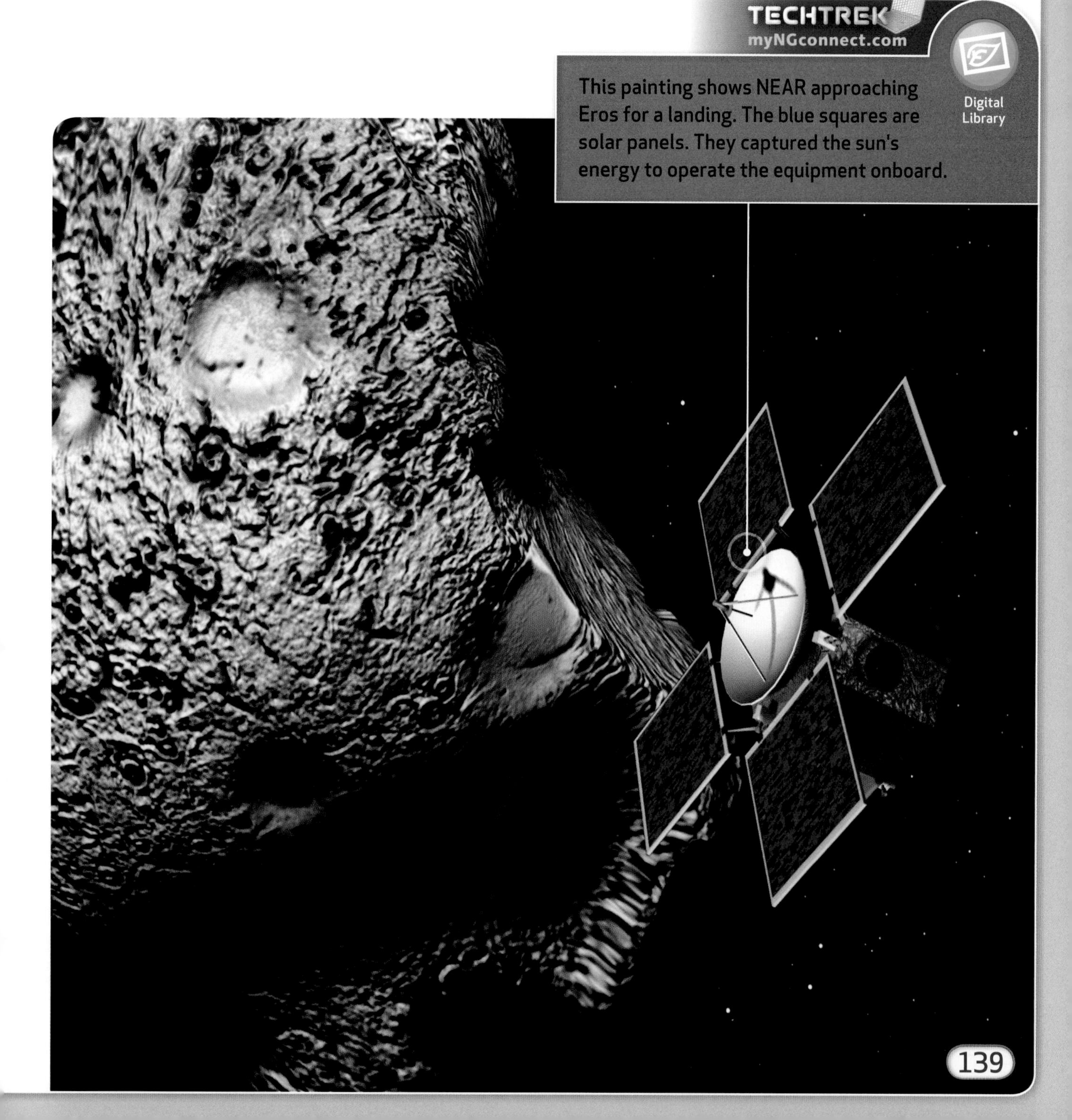

Near Earth Objects Most asteroids in the asteroid belt are not a threat to Earth. They stay in the belt for the most part. Some, however, do drift out of the asteroid belt and come closer to Earth.

Asteroids that come close to Earth are called Near-Earth Objects. The asteroid Apophis is an example of a Near Earth Object.

It was discovered just a few years ago and already scientists are watching it closely. Apophis is 320 meters (1,050 feet) wide. It passes Earth about every 16 years. With each pass, Apophis gets a little closer to Earth. In 2029, it will be only 35,000 km (22,000 miles) away.

THE SKY IS FALLING

Scientists think a large asteroid or comet hit Earth about 65 million years ago. These paintings show what might have happened.

All was peaceful a few seconds before the crash.

The asteroid or comet streaks through Earth's atmosphere. Part of it burns up.

This may seem like a safe distance, but it's close enough that Earth's gravity might change the asteroid's orbit. That could be a big problem. Instead of passing by Earth, this asteroid could eventually slam into Earth.

Today an entire program at NASA, the Near Earth Object program, is dedicated to identifying and tracking Near Earth Objects.

This crater in western Australia was formed about 300,000 years ago. It's 880 meters (2,887 feet) across. It used to be twice as deep, but windblown sand has filled it in.

The explosion sends dust into the air, and creates huge waves.

A thousand years later, only a crater remains, partly hidden underwater.

Taking Action In order to learn more about Apophis and the possibility of it impacting Earth, NASA plans to send a spacecraft to the asteroid. The spacecraft would put a homing beacon on the asteroid so that scientists could track it.

Chances are Apophis will never slam into Earth. But just in case, scientists are working on a plan. First, they want to chart Apophis' path through space. Then, they want to send another spacecraft there to collect data. The data will help them plan what to do next.

Scientists might use what they know about gravity to pull the asteroid out of its orbit. One idea is to send a special spacecraft to the asteroid. The spacecraft and its gravity would act like a tractor and pull the asteroid into a new path away from Earth.

Using powerful telescopes, scientists have discovered asteroid belts orbiting distant **stars**. Asteroids and comets are just a few of the many amazing objects in the **universe**. They are worth watching in more ways than one.

How do you move an asteroid? This painting shows the spacecraft's gravity pulling the asteroid to a new position.

universe

The universe is everything that exists throughout space.

star

A star is a ball of hot gases that gives off light and other types of energy.

NATIONAL GEOGRAPHIC

CHAPTER 3 SHARE AND COMPARE

Turn and Talk Why is it important to track the paths of asteroids and comets? Form a complete answer to this question together with a partner.

Read Select two pages in this section that are the most interesting to you. Practice reading the pages so you can read them smoothly. Then read them aloud to a partner or small group. Talk about why the pages are interesting.

my SCIENCE notebook

Write Write a conclusion that tells the important ideas about what you have learned about asteroids and comets hitting Earth. State what you think is the Big Idea of this section. Share what you wrote with a classmate. Compare your conclusions. Did your classmate recall what happens to most asteroids that enter Earth's atmosphere?

my SCIENCE notebook

Draw Look again at the Barringer Meteor Crater at the beginning of this section. Work in small groups. Draw what might have happened to make this crater. Have each person in the group draw one step in the process of how this crater formed. Compare your drawings with other groups.

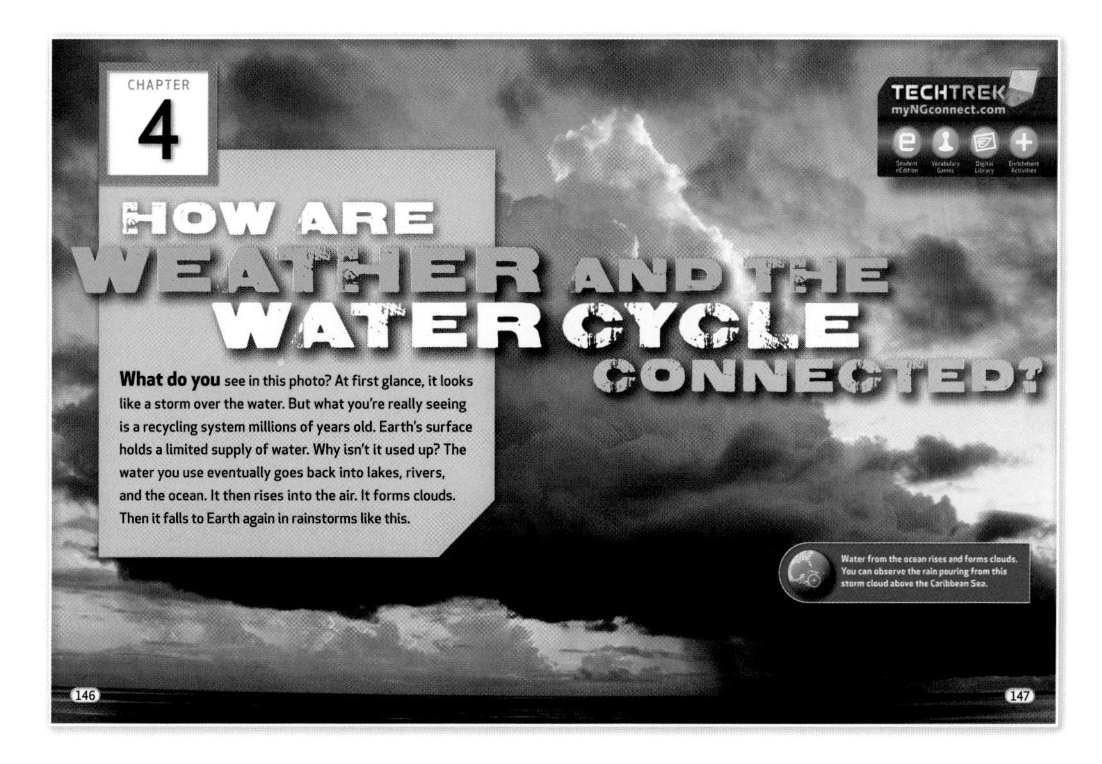

In Chapter 4, you will learn:

FLORIDA NEXT GENERATION SUNSHINE STATE STANDARDS

SC.5.E.7.1 Create a model to explain the parts of the water cycle. Water can be a gas, a liquid, or a solid and can go back and forth from one state to another. THE AIR AROUND US, THE WATER CYCLE

SC.5.E.7.2 Recognize that the ocean is an integral part of the water cycle and is connected to all of Earth's water reservoirs via evaporation and precipitation processes. THE WATER CYCLE

SC.5.E.7.3 Recognize how air temperature, barometric pressure, humidity, wind speed and direction, and precipitation determine the weather in a particular place and time. WEATHER, OBSERVING WEATHER PATTERNS

SC.5.E.7.4 Distinguish among the various forms of precipitation (rain, snow, sleet, and hail), making connections to the weather in a particular place and time. THE AIR AROUND US. THE WATER CYCLE

SC.5.E.7.5 Recognize that some of the weather-related differences, such as temperature and humidity, are found among different environments, such as swamps, deserts, and mountains. CLIMATE

SC.5.E.7.6 Describe characteristics (temperature and precipitation) of different climate zones as they relate to latitude, elevation, and proximity to bodies of water. CLIMATE

SC.5.E.7.7 Design a family preparedness plan for natural disasters and identify the reasons for having such a plan. PREPARING FOR SEVERE WEATHER

SC.5.E.7.1 Science in a Snap! Create a model to explain the parts of the water cycle. Water can be a gas, a liquid, or a solid and can go back and forth from one state to another

CHAPTER 4

HOW ARE WEATHER & WATER CONNECTED?

What do you see in this photo? At first glance, it looks like a storm over the water. But what you're really seeing is a recycling system millions of years old. Earth's surface holds a limited supply of water. Why isn't it used up? The water you use eventually goes back into lakes, rivers, and the ocean. It then rises into the air. It forms clouds. Then it falls to Earth again in rainstorms like this.

TECHTREK
myNGconnect.com

Student eEdition

Vocabulary Games

Digital Library

Enrichment Activities

AND THE CYCLE CONNECTED?

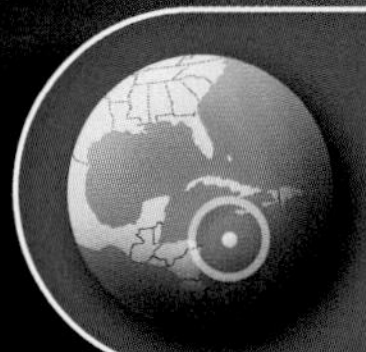

Water from the ocean rises and forms clouds. You can observe the rain pouring from this storm cloud above the Caribbean Sea.

SCIENCE VOCABULARY

weather (WE-thur)

Weather is the state of the atmosphere at a certain place and time. (p. 152)

The weather is clear and sunny.

humidity (hyū-MID-it-ē)

Humidity is the amount of water vapor in the air. (p. 153)

Humidity is a main factor that determines weather.

water cycle (WAH-tur SĪ-cul)

The **water cycle** is the constant movement of Earth's water from the surface to the atmosphere and back again. (p. 156)

In the water cycle, Earth's limited supply of water is recycled.

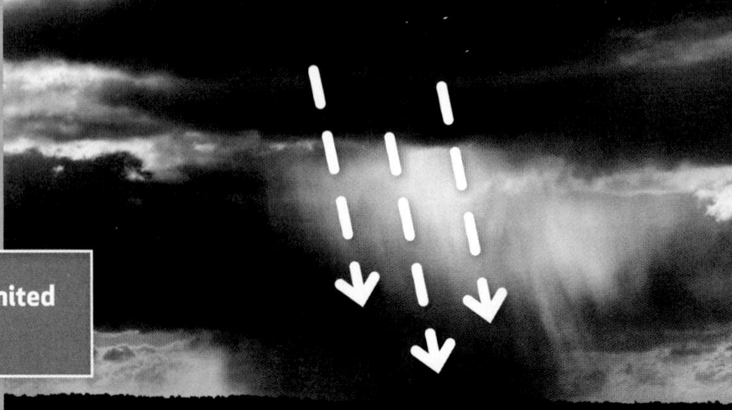

climate
(CLĪ-mit)

evaporation
(ē-va-por-Ā-shun)

front
(FRUNT)

humidity
(hyū-MID-it-ē)

water cycle
(WAH-tur SĪ-cul)

weather
(WE-thur)

evaporation
(ē-va-por-Ā-shun)

Evaporation is a change from the liquid to the gaseous state. (p. 157)

The sun causes evaporation from the water on Earth's surface.

front (FRUNT)

A **front** is the boundary where two different air masses meet. (p. 168)

Different types of fronts bring different types of weather.

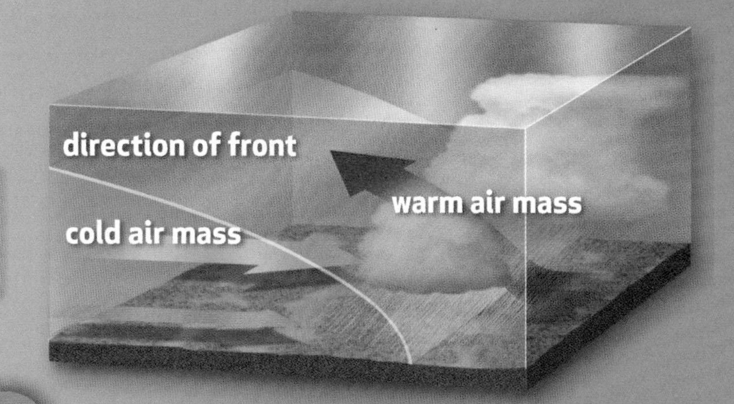

climate (CLĪ-mit)

Climate is the pattern of weather over a long period of time. (p. 170)

The climate of an area determines the plants and animals that live there.

The Air Around Us

The atmosphere is a layer of gases that wrap around Earth like a huge blanket. It holds in Earth's heat, making the planet warm enough for life. The atmosphere contains the air you breathe. It protects living things from the sun's harmful radiation. Earth's atmosphere also helps the planet recycle its water supply.

The atmosphere is made mostly of the gases nitrogen and oxygen. Earth is the only planet in the solar system with an atmosphere formed mostly of these two gasses.

Earth's atmosphere also contains countless tiny particles that float in the air. They include soil, pollen grains, and soot from fires. These particles are important for the formation of clouds.

Earth's atmosphere is about 600 km (373 miles) thick. These skydivers are falling through the lowest part of it.

Two other gases play important roles in the atmosphere, although they exist in small amounts. They are carbon dioxide and water vapor.

Carbon dioxide makes up only a fraction of one percent of the atmosphere. But it is a greenhouse gas. That means it is one of the gases that absorbs Earth's heat to keep the planet warm.

Water vapor is a green house gas, too. Water vapor is important because it plays a major role in weather. Without it, there would be no clouds. Without clouds, there would be no precipitation. Earth would be dry and lifeless.

Earth's atmosphere is 99 percent nitrogen and oxygen. The other one percent is mostly argon, with traces of other gases such as carbon dioxide, neon, helium, and water vapor.

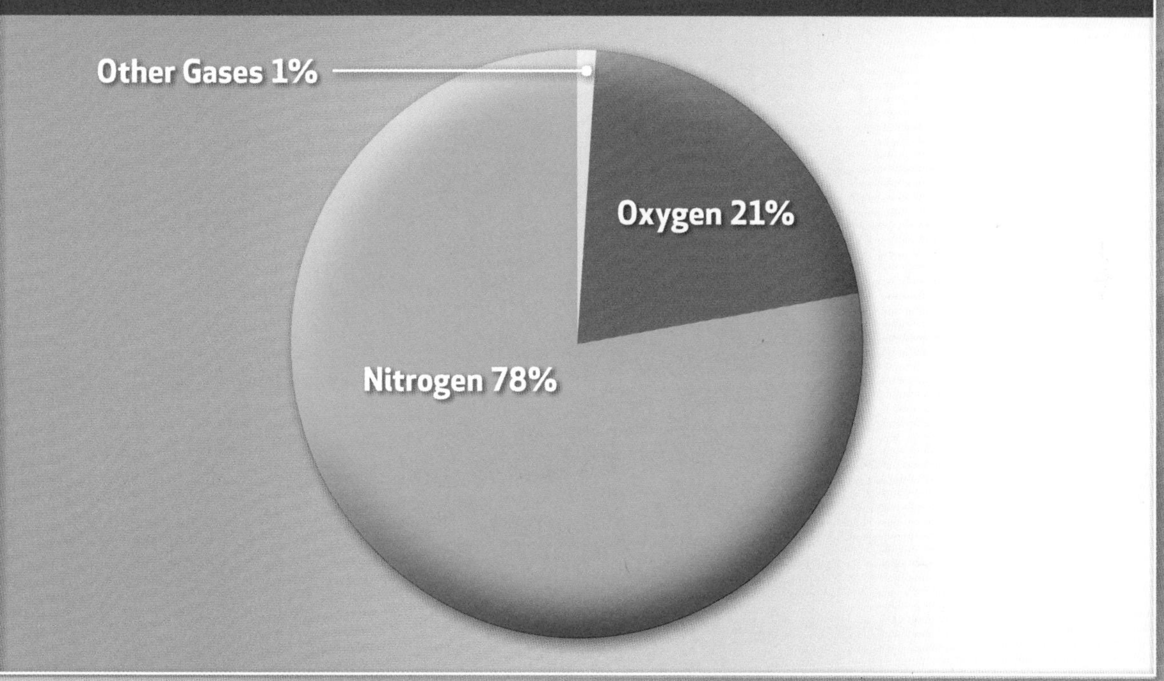

Before You Move On

1. What is the atmosphere?
2. What gas makes up most of Earth's atmosphere?
3. **Evaluate** What are some ways the atmosphere makes life on Earth possible?

Weather

When you get up in the morning, you probably check the day's **weather**. Weather is the state of the atmosphere at a certain place and time. Weather changes from season to season. It changes from day to day. Weather can even change from hour to hour. The sky might be bright and sunny in the morning, but turn rainy by afternoon.

Clouds must be at a temperature colder than 0°C (32°F) for snowflakes to form.

We can observe and describe weather by measuring properties of the air. These properties are temperature, humidity, air pressure, and wind. We can also use measurements of these properties to predict weather in the future.

Temperature Look at the photo on this page. Based on what you see, what can you infer about the temperature? There is a lot of snow on the ground and on the trees. The man in this picture is dressed in heavy clothing. The weather must be cold. Temperature is how hot or cold something is.

Humidity On hot and sticky summer days the air feels humid. **Humidity** is the amount of water vapor air can hold. Air that is humid holds a lot of water vapor. It's no surprise that air often feels wetter in summer. Warm air holds more water vapor than cool air. Humidity affects the weather.

Humid air is more likely to produce clouds and precipitation than drier air. Dry air is more likely to come with fair weather.

Observe the weather in the photo below. Do you think the humidity is low or high? The rain is a clue that a lot of moisture is in the air. The weather on this day is hot and humid, so the humidity is high.

The girl in this photo is on Lombok Island, in Indonesia. Lombok Island has warm and humid weather all year round.

Air Pressure Imagine the water pressure you feel at the bottom of a swimming pool. Earth's atmosphere is similar. Like water, air has weight. You live at the bottom of Earth's atmosphere, so there is a lot of air above you. The force of the weight of the air pressing down on you is air pressure.

Air pressure is not the same everywhere. Sometimes air pressure is low. At other times it can be high.

Lows Temperature affects air pressure. When air gets warmer, gas particles get farther apart. The air gets lighter and rises. Rising air presses down with less force, which creates a low. A low is an area of low air pressure.

The weather here is clear, with a cloudless sky Air pressure here is likely rising.

Highs When air gets cooler, the gas particles get closer together. The air gets heavier and sinks. It presses down with more force, creating a high. A high is an area of high pressure.

Clouds can't form where air is sinking. So when air pressure gets higher you will often see clear skies. On the other hand, when a low forms, clouds tend to form, and you will often have rain.

Wind

Wind is the movement of air from areas of high pressure to areas of low pressure. It's like water running down a hill. When the difference in air pressure is great, wind blows faster. When the difference in pressure is small, wind is slower.

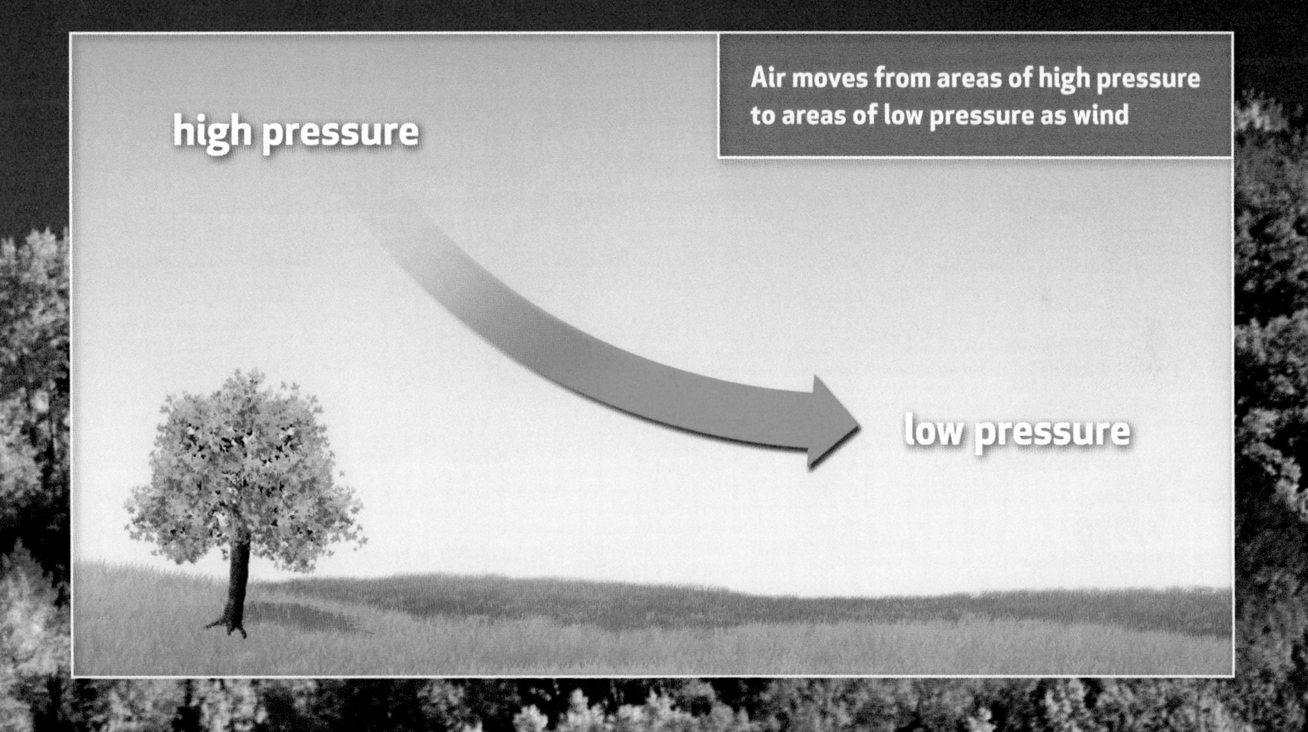

Air moves from areas of high pressure to areas of low pressure as wind

Before You Move On

1. What does air feel like when humidity is high? Why does it feel that way?
2. How does temperature determine air pressure?
3. **Draw Conclusions** Why does low air pressure often mean cloudy and rainy weather?

The Water Cycle

Water covers more than 70 percent of Earth's surface. Most of Earth's water is in the salt water ocean that covers most of the planet. But some of Earth's water is also the fresh water , water with little salt, in rivers and lakes. Some fresh water is frozen as glaciers and ice caps. Groundwater is fresh water below the surface. It collects as rain seeps into soil, filling spaces between soil and rock.

Humans use a lot of water. But Earth's supply never runs out because the **water cycle** is at work. The water cycle is the constant movement of water from Earth's surface to the atmosphere and back again. Through the water cycle all of Earth's bodies of water are connected to one another. Salt water can become fresh water, and fresh water can become salt water. Because of the water cycle, Earth's limited water supply has lasted for billions of years.

The Montauk Point Lighthouse in Long Island, New York, overlooks the Atlantic Ocean. Water evaporates from the Atlantic Ocean and enters into the atmosphere.

Evaporation When the sun shines on the ocean, it heats the water. The warmer temperature makes the liquid water change to water vapor, a gas. This change from a liquid to a gas is called **evaporation**.

What happens when water evaporates? Particles of liquid move from the surface of the water to the air above. Water particles are always crashing into each other. When the water gets warmer, the particles gain energy. They crash more often. More water particles escape from the water's surface and enter into the air.

The warmer water is, the faster it evaporates. Evaporation is how water enters Earth's atmosphere. Earth's ocean is the largest source of water for evaporation.

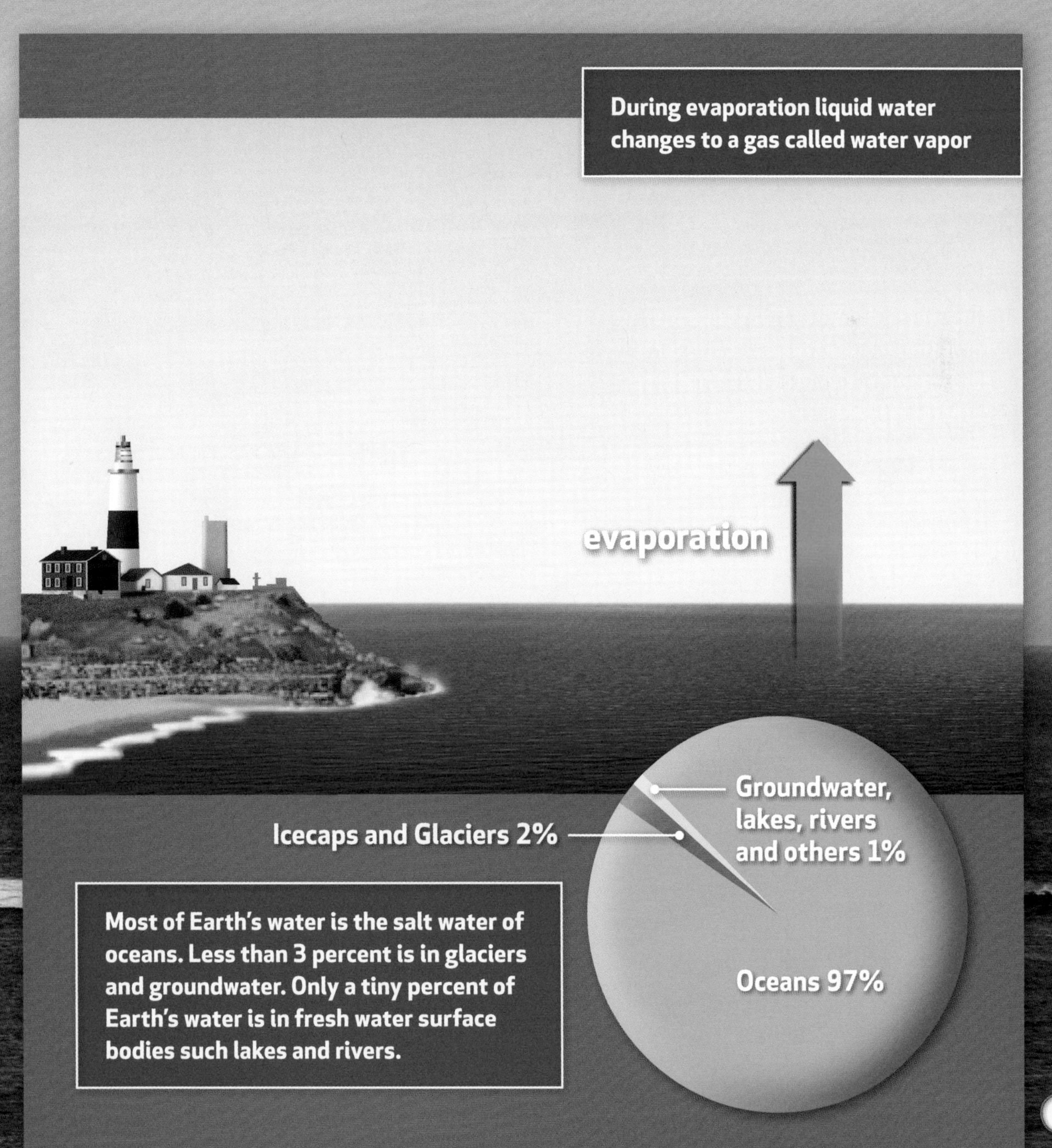

Condensation Early in the morning, you may see tiny drops of water on leaves or blades of grass. This is dew, which is caused when water vapor in the air changes into water droplets. Fog and clouds form from water vapor, too.

When water vapor in the air cools, it condenses. Condensation is the change from a gas to a liquid. In the lower atmosphere, air gets cooler the higher it goes. As air rises, the water vapor condenses into clouds.

When air pressure lowers and warm air rises, clouds often form and rain falls. How does this happen? Warm, humid air near the ground rises. When the water vapor in the air reaches cooler air above, the water vapor can condense.

Tiny water droplets float together to form these huge clouds. It the air is cold enough , the droplets freeze to form ice crystals

condensation

Just as water vapor condenses on a surface to form dew, a surface is also needed to form clouds. There are millions of tiny bits of dust in the air. Each tiny speck provides a surface on which water vapor can condense. When water vapor condenses around these specks cloud droplets form.

Cloud droplets are tiny. They are much smaller than rain drops. In fact, it would take one million tiny cloud droplets to make just one drop of rain. The cloud droplets are also very light. So they float on the air in the atmosphere.

Precipitation Rain, snow, sleet, freezing rain, and hail are types of precipitation that fall from clouds. The type of precipitation depends on the temperature of clouds and the temperature of the air between the clouds and the ground.

Rain Rain often starts in clouds as snow. If the snow falls though air that is above freezing on its way to the ground, it melts. It then hits the ground as drops of rain.

Snow Snow starts in clouds as snow. Unlike rain, snow falls through a layer of air that is below freezing. So it remains snow.

Sleet Sleet starts as snow, too. The snow melts into rain as it falls through a layer of warmer air below the clouds. If there is a layer of freezing air between the warm air and the ground, the rain freezes. It then falls as tiny balls of ice called sleet.

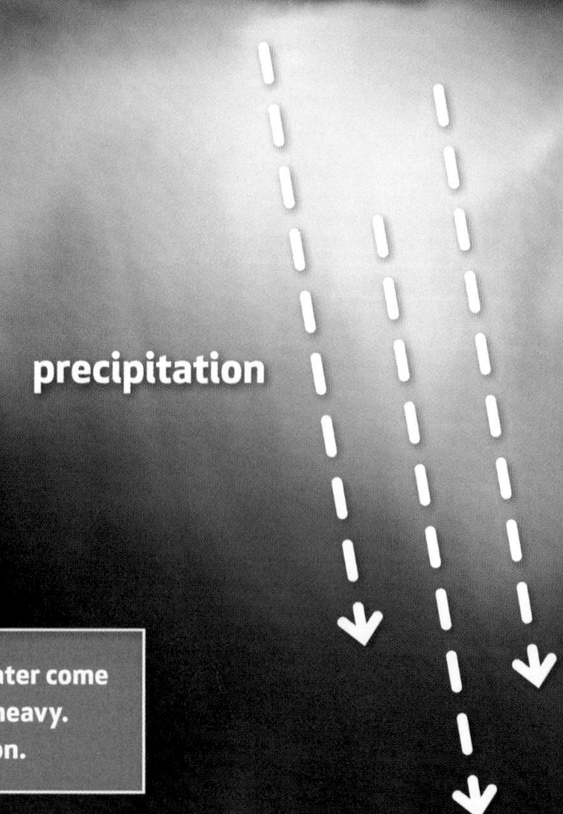

When many droplets or crystals of water come together in the clouds, they become heavy. Then they fall to Earth as precipitation.

Freezing Rain Freezing rain forms in almost the same conditions as sleet. But the layer of freezing air near the ground is thinner. When the rain falls through it, there isn't enough time for the water to freeze. Instead, it becomes supercooled. Supercooled water is liquid even though its temperature is below freezing. When this super cold water hits a tree branch or the ground, it turns into a layer of ice.

Hailstones Hailstones are round lumps of ice. They start as small ice pieces in storm clouds. Winds in the clouds bounce them up and down through cloud droplets. The cloud droplets freeze onto the ice, making the hailstones bigger and bigger. When the hailstones are heavy enough, they fall from the cloud.

Water that falls to Earth's surface is called precipitation.

The water cycle is completed as water returns to Earth's surface as precipitation. When the water reaches the ground, some falls back into surface water bodies, such as oceans, rivers, and lakes.

Water that falls on land might become runoff. Runoff is water that does not sink into the soil. It flows over the surface of soil, streets, and sidewalks before returning to streams, lakes, or the ocean.

Water can seep into soil as well. It then trickles down and becomes part of Earth's supply of underground water. Groundwater flows slowly. Still, it eventually returns to surface waters such as rivers and lakes through underground pathways. It can then evaporate and make its trip through the water cycle again.

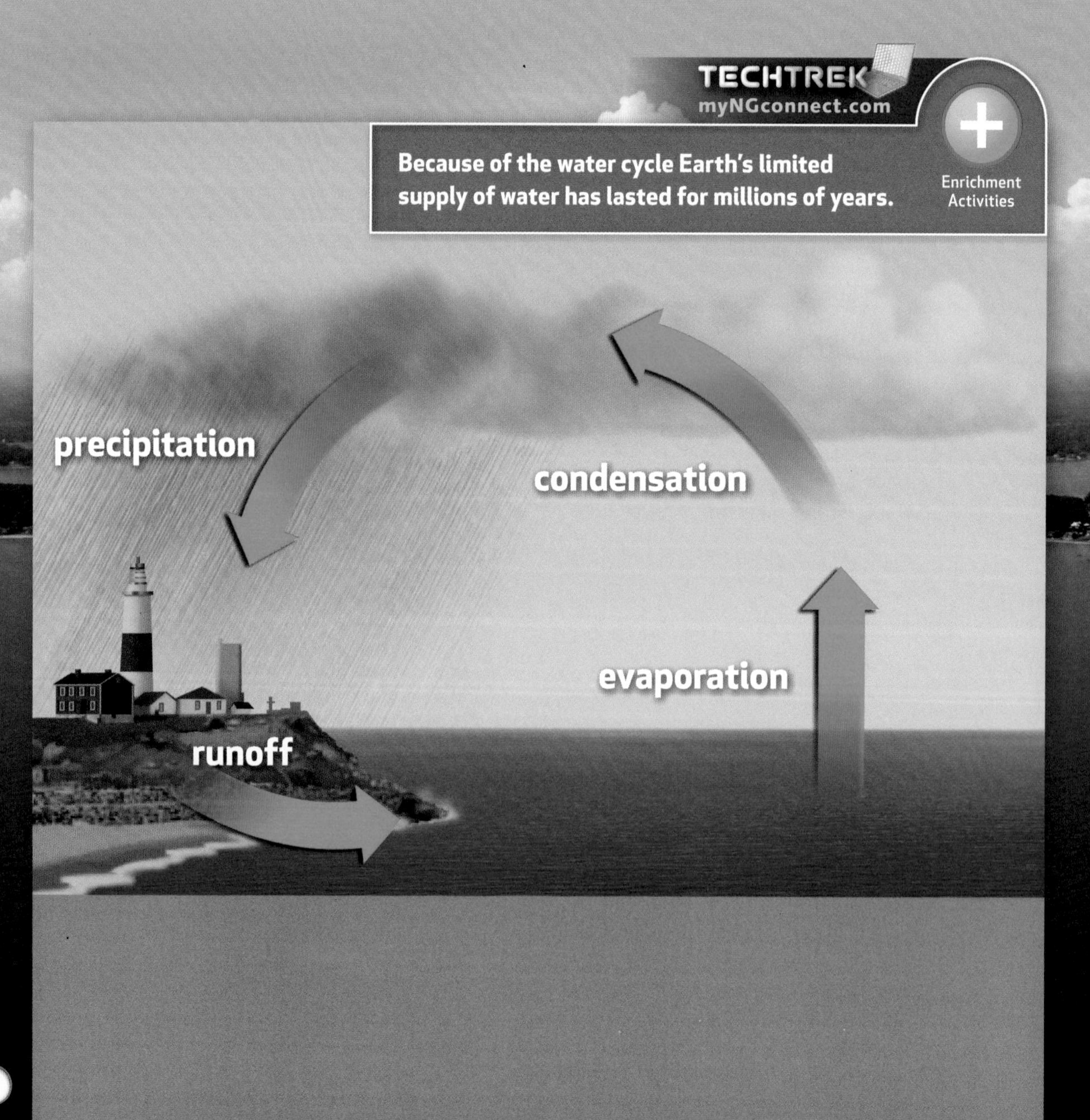

Science in a Snap! Groundwater

Put a layer of gravel or small rocks at the bottom of a clear, plastic container. Cover the gravel with a three-inch layer of sand.

Poke small holes into the bottom of a paper cup with a paper clip. Sprinkle water gently over the sand. This models rain falling on soil. Observe what happens to the water.

How does this show what happens to rain that seeps into the ground?

Before You Move On

1. What is the water cycle?
2. Contrast evaporation and condensation.
3. **Evaluate** Why is the water cycle important for Earth?

Observing Weather Patterns

Scientists observe and study weather patterns to predict the weather in the future. To predict the weather, they use different weather instruments to collect data about the current weather. The chart below shows some of these instruments.

Each instrument in the chart measures just one specific part of the weather, such as temperature. Scientists make their predictions by combining data from many different instruments.

THERMOMETER

Thermometers measure the temperature in degrees. This helps decide how the air will feel and predict weather that could be unsafe.

ANEMOMETER

Anemometers measure wind speed in kilometers or miles per hour. Knowing how fast the wind is blowing can help figure out how much damage a storm may cause.

WIND VANE

Wind vanes measure the direction the wind is blowing: north, east, south, or west. A change in wind direction can mean a change in weather.

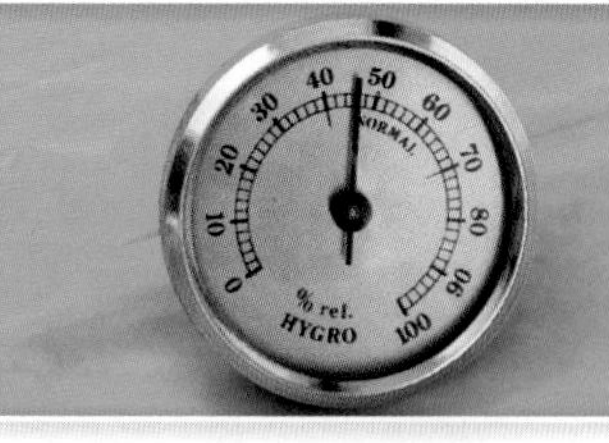

HYGROMETER

Hygrometers measure the humidity in the air. Hygrometer readings can show when precipitation may happen, or how uncomfortable the air will feel.

BAROMETER

Barometers measure air pressure. It is also called barometric pressure. Increasing air pressure can mean fair weather. Decreasing air pressure can mean rainy weather.

RAIN GAUGE

Rain gauges measure rainfall in centimeters or inches. Scientists can use rainfall measurements to help predict droughts or floods.

Scientists use other tools, too. Weather satellites show the movement of storms. Weather balloons rise into the sky all over the world each day. They collect data on temperature, air pressure, and humidity. Scientists also use several types of radar to "see" inside clouds. Using radar can help them find the strength of storms. They can also find clouds that might produce dangerous storms such as tornadoes.

Weather data from thousands of sources goes into computers. The computers hold models that help scientists make forecasts, or predictions. Scientists also make weather maps. The maps show where storms are moving, as well as areas of high and low air pressure.

These scientists are collecting data at a weather station at Racer Rock in Antarctica. This weather station measures wind, temperature, air pressure, and humidity.

Air Masses Bring Weather

What's the weather outside? Chances are it's generally the same in a large area surrounding your home. Air moves over Earth's surface as huge masses. In fact, most air masses are more than 1600 kilometers (1000 miles) across and several kilometers thick. The temperature and humidity of the air is about the same throughout an air mass. This usually means that the weather, in an air mass, is the same.

But a neighboring air mass may have a different temperature and humidity, and different weather. Knowing what kind of air mass is on the way can help you predict what kind of weather will occur in the near future.

Every air mass takes on the properties of the area where it formed. Some form in cold regions near the poles. These are cold air masses. Others form in warm areas closer to the Equator. Some air masses form over land and others form over water.

Scientists all over the world launch weather balloons twice each day. The balloons carry instruments that gather data at various heights in the atmosphere.

When air masses move they affect the weather of places they pass over. Look at the map. It shows that at any given time, several air masses affect the weather in the United States.

When air masses form over Canada, they are cold and dry. These air masses pass over the northern and central part of the U.S. These parts of the country have cold, dry weather in winter.

Air masses that form over cold polar water bring cold and wet weather to northern coasts. Air masses that form over warm ocean water near the equator bring warm and humid air.

When air masses form over Mexico, they are warm and dry. This is why the Southwest has hot, dry weather.

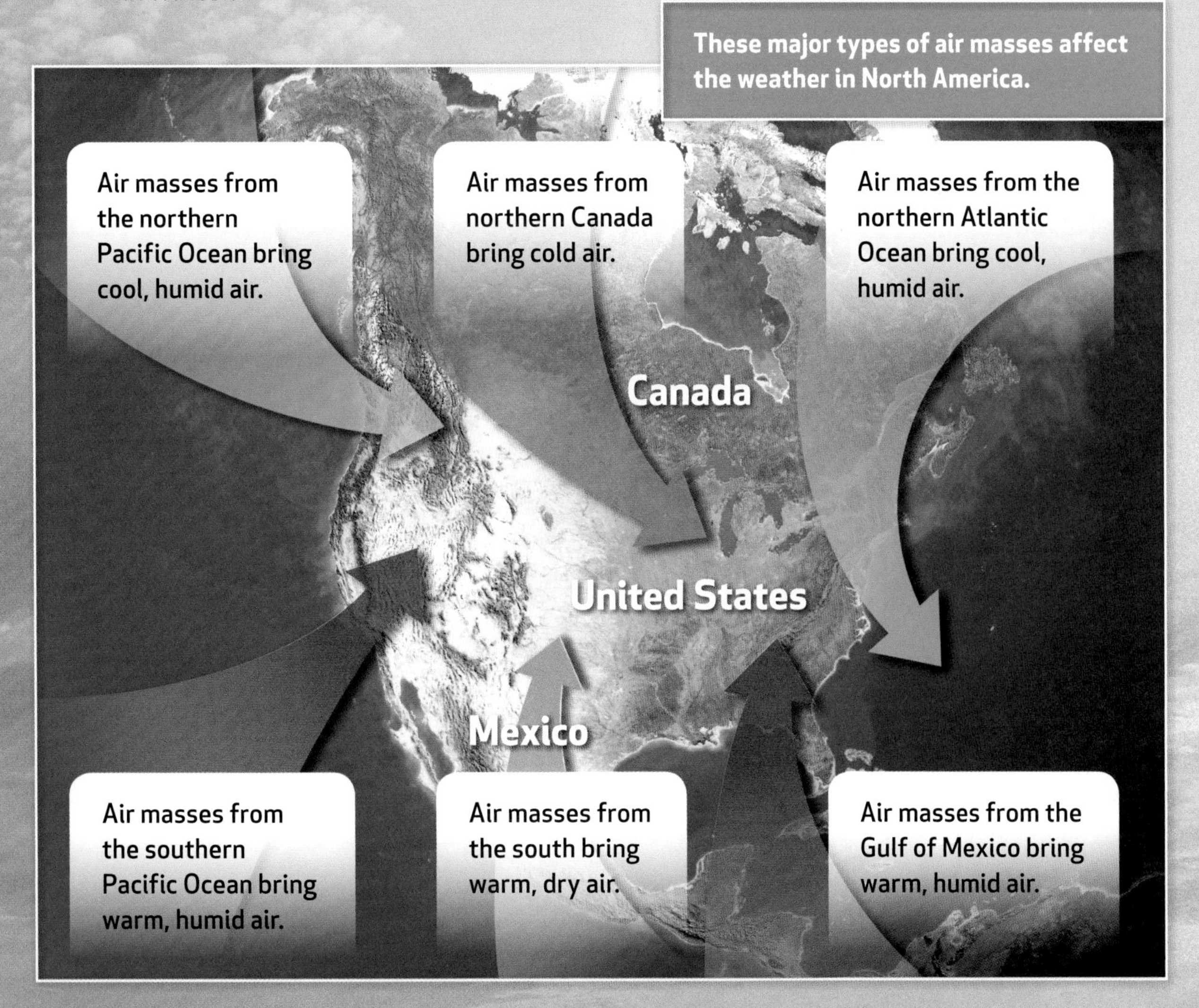

These major types of air masses affect the weather in North America.

Fronts Remember that an air mass has the same temperature and humidity throughout. Some air masses are made up of warm, moist air and some contain cooler, drier air. What happens when two air masses meet? The weather changes. Thunderstorms, rain showers, strong winds and other weather events happen at the boundary between air masses.

The leading edge of a air mass is called a **front**. The edge of a moving mass of cooler air is a cold front. The edge of a moving mass of warmer air is a warm front

Warm Fronts Warm fronts bring warmer temperatures. At warm fronts, a faster-moving warm air mass meets a cold air mass. The faster, lighter warm air rides up over the heavier cold air. As the warm air rises and cools, the water vapor in the air condenses to form clouds.

At a warm front, the warmer air rises at a low angle. It is like walking up a gradual hill instead of climbing a steep wall. The warm air covers a larger area. The sky fills with thicker, lower clouds and there can be light rain that lasts for hours or days.

This satellite image shows a front moving over the north Atlantic Ocean.

Cold Fronts Cold fronts bring cooler temperatures. Cold fronts form where a faster-moving cold air mass meets a warm air mass. The faster, heavier cold air plows under the warm air. It pushes the warm air up sharply. Water vapor in the rising air condenses to form clouds.

At a cold front, the warmer air rises steeply. It is more like climbing a steep wall than walking up a gradual hill. The rising air cools faster. It has less area to spread out. Tall storm clouds, called cumulonimbus clouds, may form. These bring short but heavy showers. Sometimes cold fronts bring severe weather, including tornadoes.

Air masses and fronts bring certain types of weather. Scientists can predict this weather by knowing where air masses formed and the track they will take.

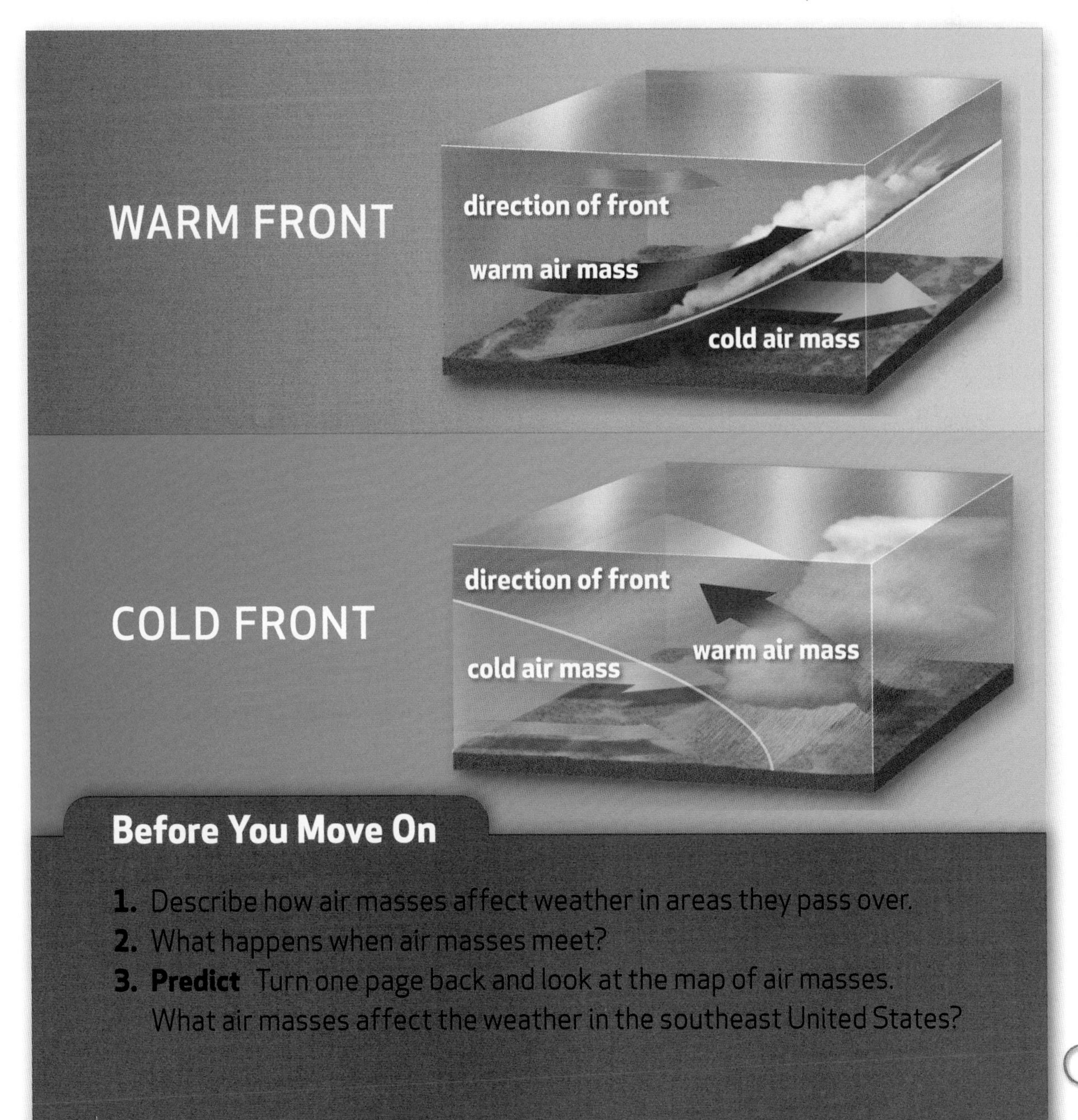

Before You Move On

1. Describe how air masses affect weather in areas they pass over.
2. What happens when air masses meet?
3. **Predict** Turn one page back and look at the map of air masses. What air masses affect the weather in the southeast United States?

Climate

When someone says "It is hot and humid here in the summer," they are talking about the **climate** of a particular place. Climate is the general weather of an area over a long period of time. Some places have warmer climates than others. Florida has long, hot summers and warm winters. Alaska has short, cool summers and cold winters.

The climate in Florida is different from the climate in Alaska.

Look at the map of the climate zones of the United States. The range of temperatures and the amount of precipitation a place experiences determines its climate. Each type of climate has certain weather patterns that are the same or repeated year after year.

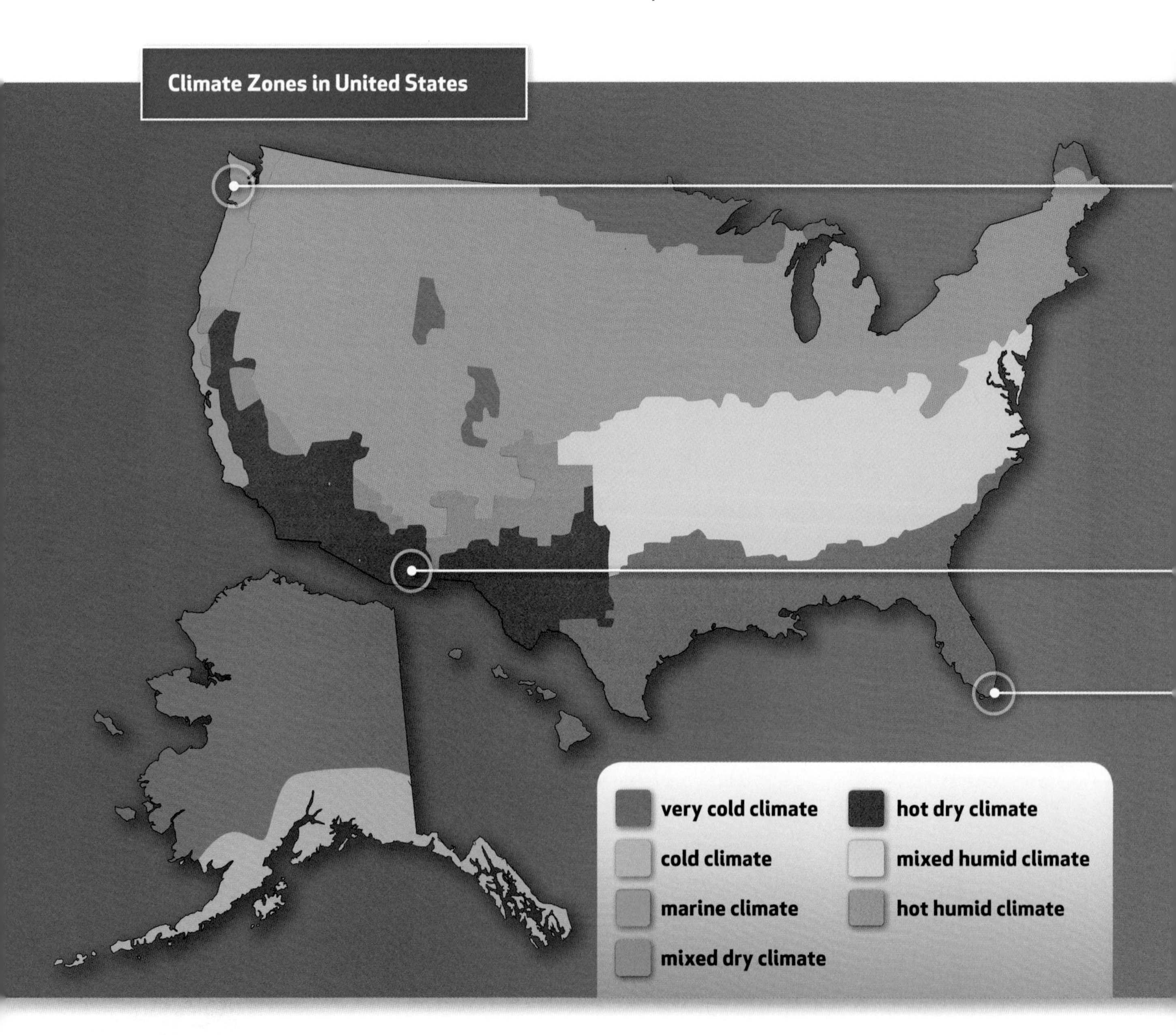

The climate of a region also affects the environment found there. Deserts are usually very dry, but they can be either cold or hot. Swamps usually occur in warm and wet environments.

Mountains can have different climates. The top of a mountain usually have cool temperatures and low humidity. But the base of the mountain will have the climate of the surrounding area.

The mountains in Washington State are in a marine climate at the base of the mountains. The temperature goes down as you climb higher.

Picacho Peak State Park near Tuscon, Arizona has a hot, dry climate. The weather in this hot desert is very dry. Less than 25 centimeters (about 10 inches) of rain fall each year.

Southeasten Florida has a hot and humid climate. The swamps of the Everglades contribute to the warm and humid conditions.

Factors Affecting Climate

Average temperature and precipitation are the key parts of every climate. What determines a climate's temperature and precipitation? Latitude, elevation, and nearness to large bodies of water are all factors that can affect climate.

Latitude Latitude is the distance north or south of the Equator. Latitude affects temperature. Generally it is warmest near the Equator and coolest near the poles. So the closer an area is to the Equator, the warmer its climate.

Elevation Elevation affects climate, too. Look at the photo. The meadow is green. Why do you think the top of the nearby mountain is covered in snow?

Recall that air gets cooler as it rises. Places that are elevated have cooler temperatures than nearby lowlands. This mountain rises several kilometers above the surrounding land. The temperature on the mountain is much cooler.

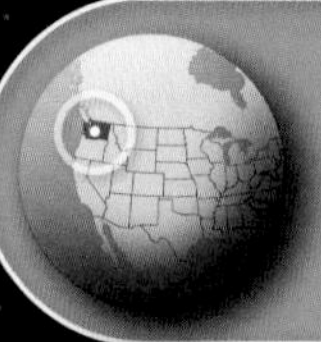

Ice and snow cover the top of Mt. Rainier even in summer.

Find the Equator on the map. Earth's three major temperature zones are based on latitude.

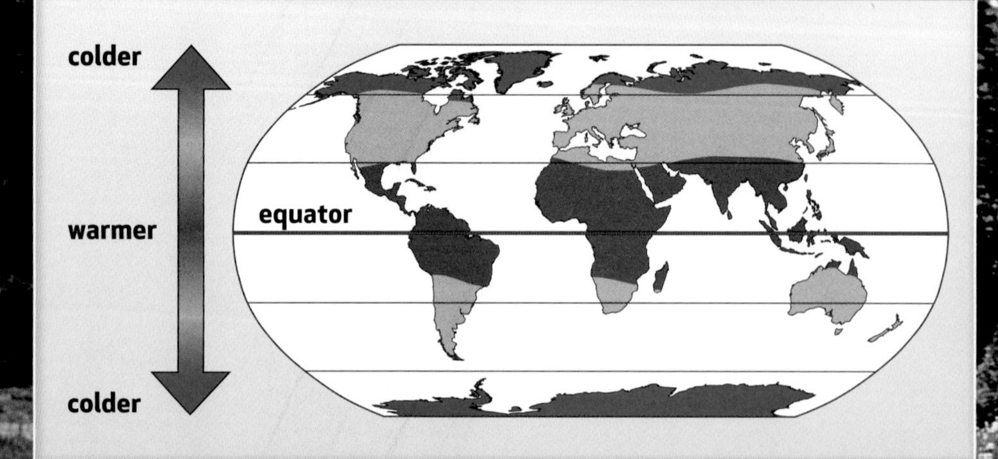

Large Bodies of Water

Temperatures are usually milder near large bodies of water such as oceans and large lakes. Water heats and cools at a slower rate than land. Also, the water does not get as hot or as cold as the land does. How does this affect climate? For example, ocean water heats up during the summer. During the fall the ocean water slowly releases this heat and warms the air. This means that temperatures near the coast may not be as cold as temperatures inland away from the water.

FACTORS THAT AFFECT **TEMPERATURE AND PRECIPITATION**

FACTOR	ITS EFFECT
LATITUDE	Areas nearest the equator are warmest. Temperature generally decreases toward the poles.
ELEVATION	Climate cools with increasing elevation.
NEARNESS TO LARGE BODIES OF WATER	Large bodies of water moderate the climate of nearby land. Marine climates are milder and have a smaller range of temperature than continental climates.

Comparing Climates Here you can observe two cities with different climates. They are Fargo, North Dakota, and Key West, Florida.

First observe the photo and the data on Fargo. The city is in the northern plains. It is farther from the Equator than Key West. There are no mountains or large bodies of water nearby.

What can you infer about Fargo's climate? Look at the chart on this page. You can see that the hottest month is July and the coldest month is January.

You can see that Fargo's climate is cold and dry in the winter and warmer and more humid in the summer.

Climate Data for Fargo, North Dakota (2008)

MONTH	AVERAGE HIGH TEMPERATURE °C	(°F)	AVERAGE PRECIPITATION cm	(in)
JANUARY	-8.9	(15.9)	2.0	(0.8)
FEBRUARY	-5.1	(22.8)	1.5	(0.6)
MARCH	1.8	(35.3)	3.1	(1.2)
APRIL	12.5	(54.5)	3.6	(1.4)
MAY	20.8	(69.5)	6.6	(2.6)
JUNE	25.2	(77.4)	8.9	(3.5)
JULY	27.9	(82.2)	7.4	(2.9)
AUGUST	27.2	(81.0)	6.4	(2.5)
SEPTEMBER	21.1	(69.9)	5.6	(2.2)
OCTOBER	13.4	(56.1)	5.1	(2.0)
NOVEMBER	1.8	(35.2)	2.8	(1.1)
DECEMBER	-6.2	(20.8)	1.5	(0.6)

Fargo is in the very cold climate zone.

Now observe the data for Key West. Key West is an island at the tip of Florida. Key West is closer to the Equator than Fargo. Warm and humid air masses move over it throughout the year. This, along with the warm ocean waters give Key West a warm and humid climate.

You can also see that Key West is wetter than Fargo. Summers are hot and very humid, with frequent afternoon thunderstorms.

Climate Data for Key West, Florida (2008)

MONTH	AVERAGE HIGH TEMPERATURE °C	(°F)	AVERAGE PRECIPITATION cm	(in)
JANUARY	24.1	(75.3)	5.6	(2.2)
FEBRUARY	24.4	(75.9)	3.8	(1.5)
MARCH	26.0	(78.8)	4.8	(1.9)
APRIL	27.7	(81.9)	5.3	(2.1)
MAY	29.7	(85.4)	8.9	(3.5)
JUNE	31.2	(88.1)	11.7	(4.6)
JULY	31.9	(89.4)	8.4	(3.3)
AUGUST	31.9	(89.4)	13.7	(5.4)
SEPTEMBER	31.2	(88.2)	14.0	(5.5)
OCTOBER	29.3	(84.7)	10.9	(4.3)
NOVEMBER	27.0	(80.6)	6.6	(2.6)
DECEMBER	24.8	(76.7)	5.3	(2.1)

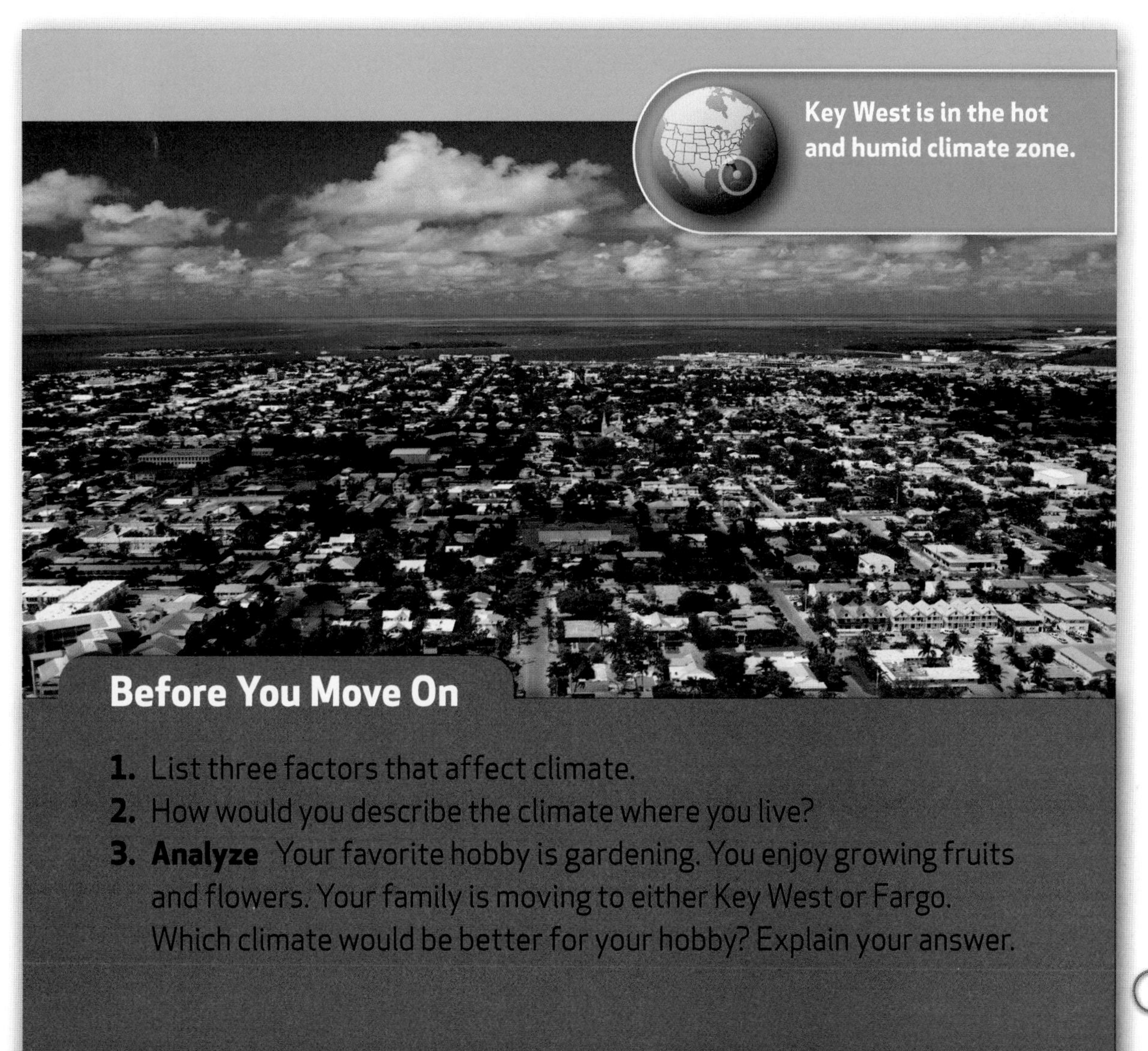

Key West is in the hot and humid climate zone.

Before You Move On

1. List three factors that affect climate.
2. How would you describe the climate where you live?
3. **Analyze** Your favorite hobby is gardening. You enjoy growing fruits and flowers. Your family is moving to either Key West or Fargo. Which climate would be better for your hobby? Explain your answer.

NATIONAL GEOGRAPHIC

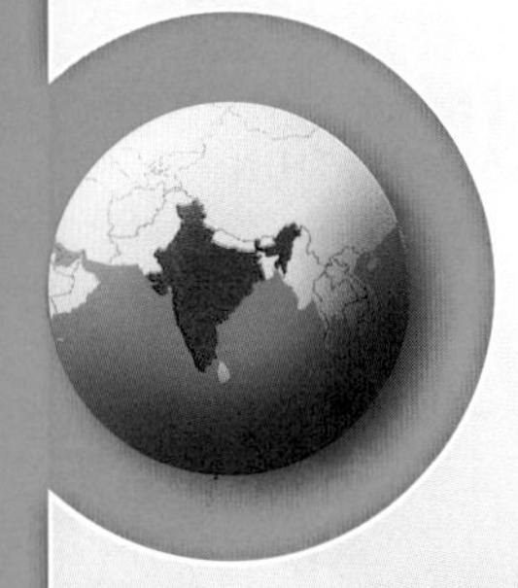

THE MONSOON

Each summer, the farmers of India wait for the monsoon. It brings rains they need for their crops. A monsoon is a seasonal change in the direction of the normal wind pattern. The term comes from the Arabic word *mausim*, meaning "season". Monsoons cause wet and dry seasons throughout much of the tropical region.

The winter monsoon brings the dry season. At this time of the year the winds blow to the south. Cool air over India and other parts of Asia pushes south toward the ocean. Because the air comes from over the land, it doesn't have much moisture. Therefore, it doesn't rain much.

In India, the winter monsoon brings clear skies and dry weather.

During the summer monsoon the wind pattern changes direction. Hot temperatures over the land heat the air. Cooler, humid air is blown in from the ocean toward the land. As the cooler, humid air moves inland, it is heated by the warm air. This causes the air to rise, drop its moisture as rain, and be replaced by more cooler humid air moving in from the ocean. This weather pattern repeats until the winds change direction in the fall.

This farmer is preparing a paddy field for rice in the heavy rain of the summer monsoon.

The summer monsoon, which normally produces heavy rains in India, carries moist ocean air over large parts of Asia. During the Northern Hemisphere winter, monsoons bring needed rainfall to Australia and Indonesia. Half the world's population depends on yearly monsoon rains to provide water for crops.

Here is the same location during the summer monsoon. Precipitation has changed the landscape.

Conclusion

Weather occurs in the atmosphere. Many factors affect the weather, including temperature, humidity, and air pressure. The water cycle is responsible for the formation of clouds and precipitation. Scientists use instruments to collect data about all of these factors. They observe weather patterns to predict what the weather will do in the future. Weather patterns over a longer period of time are climate. Climate is affected by different factors, such as latitude, elevation, or nearness to a large body of water.

Big Idea Important weather events are tied to the water cycle.

WEATHER PATTERNS AND EVENTS

THE WATER CYCLE
- evaporation
- condensation
- precipitation

WEATHER
- temperature
- humidity
- wind
- air pressure

CLIMATE
- latitude
- elevation
- nearness to large bodies of water

Vocabulary Review

Match the following terms with the correct definition.

A. weather
B. climate
C. humidity
D. evaporation
E. front
F. water cycle

1. The boundary where two air masses meet
2. The amount of water vapor in the air
3. The change in state from liquid to a gas
4. The pattern of weather over a long period of time
5. The state of the atmosphere at a certain place and time
6. The constant movement of Earth's water from the ground to the atmosphere and back again

Big Idea Review

1. **Name** Name the steps in the water cycle.
2. **Recall** What makes up the atmosphere?
3. **Explain** Why do clear skies often occur over areas of high pressure?
4. **Classify** Which of these does not belong? *hail, runoff, sleet, rain, snow*. Explain your answer.
5. **Make Judgements** The weather forecast states that an area of high pressure is moving into your area. Should you take your umbrella today or leave it at home? Explain your answer.
6. **Analyze** It's a summer day in a city with a desert climate. It's also a summer day in a city near the ocean with a warm, humid climate. How would the day be different in both places, based on the climate?

Write About Weather

Interpreting Diagrams This diagram shows a warm front. Explain what is happening. Then describe the kinds of weather you can expect from a warm front.

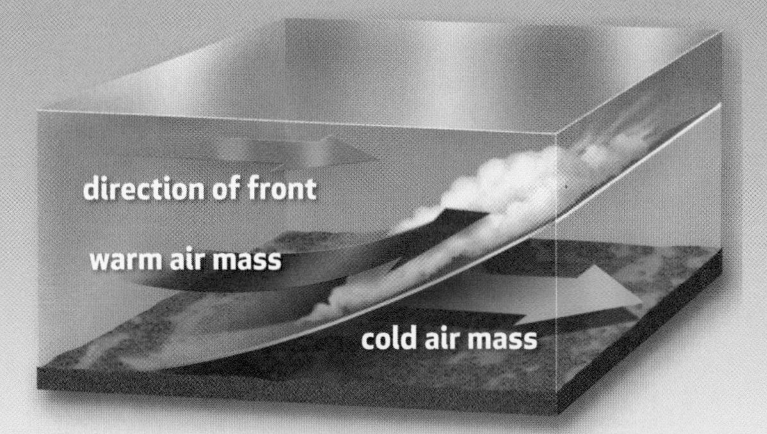

EARTH SCIENCE EXPERT: METEOROLOGIST

What's it like to look inside a storm? Ask a meteorologist.

For people like Sepi Yalda, the sky is a puzzle. How does it change—and how can we predict those changes? Yalda teaches meteorology, the study of weather and the atmosphere.

How did Yalda get interested in the atmosphere? She liked to solve puzzles. "I enjoyed experiments where I wasn't sure of the outcome," she says of what she liked most about science in school. She likes the fact that there are a lot of questions that still need answers in meteorology.

People depend on the weather predictions made by meteorologists everyday.

Every day meteorologists answer one big question for all of us. What will the weather be like? To get the answer, they collect data on temperature, winds, humidity, and air pressure. The data comes from hundreds of weather stations. It comes from weather balloons and planes that fly through storms. Satellites beam images of the atmosphere to Earth. Meteorologists put the data into computers that help them make predictions.

TECHTREK
myNGconnect.com

Student eEdition

Digital Library

Yalda is working to find better tools for gathering data and teaching meteorology. "It is always neat to be the first to test a new 3-D tool," she says of a new gadget that lets you visualize flying through winds high up in the atmosphere.

Still, meteorologists do more than predict the weather. They experiment with ways to decrease air pollution. They study the way storms form. They also provide data that helps engineers design buildings and roads, helps farmers plan crops, and warns people about floods and hurricanes.

Think you want to be a meteorologist? Like Yalda, you must earn a degree in meteorology. You will take courses in math, computer science, physics, and chemistry. Many meteorologists work for the government, universities, or on radio and TV.

Sepi Yalda has a doctorate in meteorology. She does research in meteorology and teaches at a university.

NATIONAL GEOGRAPHIC

BECOME AN EXPERT

Preparing for Severe Weather

On most days, the information from weather forecasters helps people decide what clothes to wear outdoors, or what outdoor activities to plan. But sometimes weather forecasters perform a much more important task. When weather becomes extreme, they must warn about severe and possibly dangerous weather conditions so that people can prepare and be safe.

Forecasters study data from satellites, radar, and other instruments to try to predict severe weather events. But predictions of extreme weather are not always correct. Weather can change quickly, and in unexpected ways. Ordinary weather patterns can develop into severe storms in days, hours, or even minutes. People living in a storm's path may not have time to leave the area.

Because severe weather can strike quickly, it's important to plan and prepare for it ahead of time. Planning includes knowing what to do and where to go if extreme weather hits. It's also a good idea to have an emergency kit set aside in case you are caught in a storm's path. This kit should have supplies and tools to help you stay safe.

Radar images help forecasters predict severe storms.

Most weather is not severe enough to cause harm. Yet, high winds flatten houses, floods wash out bridges, and blizzards cover buildings with snow. Severe weather doesn't happen every day. But when it does, people should be ready.

Set aside water, food, and other supplies in case of severe weather.

EMERGENCY SUPPLY KIT

Keep an emergency supply kit handy.

Include:

- Bottled water (enough for 3 days)
- Flashlight
- Battery-powered radio
- Extra batteries for radio and flashlight
- Cell phone with charger
- First aid kit
- Food that won't spoil, such as canned or dry food (enough for 3 days)
- Can opener
- Paper cups, plates, and plastic utensils
- Blankets or sleeping bags (one for each person)
- Matches in waterproof container
- Fire extinguisher
- Pet food (if you have a pet)
- Personal identification
- Credit card and cash
- Extra set of car keys
- Wrench or pliers to turn off utilities
- Whistle or signal flair to signal for help
- Map of your area
- Medications needed by family members
- Any other special items needed by infants, elderly, or disabled persons

Hurricanes In August 2004, Hurricane Charley slammed into Florida. Charley was a destructive storm. Its 233 kilometer (145 mile) per hour winds wrecked thousands of homes and ripped up trees. The storm surge, the dome of water a hurricane pushes in front of it, flooded low areas on the coast.

People cover glass windows with plywood to protect them from a hurricane's high winds.

Hurricanes are Earth's biggest storms. They are hundreds of kilometers wide, with winds that can top 250 kilometers (155 miles) per hour. Those that reach the United States form off the west coast of Africa. They develop spiral winds around a center of low pressure. Warm ocean waters and high **humidity** feed the storms. The **evaporation** of humid air from the ocean gives them energy.

When forecasters issue a Hurricane Watch, the storm is still one or two days away. There is a good chance it will hit the area. But it's not a sure thing. To protect their homes, people cover their windows with plywood.

People prepare for a hurricane by putting sand in sandbags. The sandbags will be used to block flood water from storm surges.

humidity

Humidity is the amount of water vapor in the air.

evaporation

Evaporation is a change from the liquid to the gaseous state.

They move items inside that can blow around. They check emergency kits for supplies they will need if power goes out and floodwaters block roads.

A Hurricane Warning means a hurricane will strike. Every family should have an emergency plan. It tells everyone where to go, who to call, and what to do if a storm hits.

As the storm gets closer, those who are in charge will tell people to evacuate. Many of these people live in low areas that flood. Some families go to shelters that keep people safe during the storm. Others choose to remain at home. They stay in interior rooms away from windows and doors.

FAMILY EMERGENCY PLAN INFORMATION

Have a plan to use in case of an emergency, such as a hurricane. Review the plan with your family. List the following information and keep it where you can find it.

- A place where everyone meets in an emergency
- Names and phone numbers of people to contact
- Addresses and phone numbers of family members' work locations and schools
- Name and phone number of family doctor or hospital
- Name and date of birth of each family member
- Any special needs of family members, such as medications or allergies

Hurricanes are ranked on a scale of 1 to 5, with Category 5 storms being the most powerful. This satellite photo shows Charley, a Category 4 storm.

Tornadoes Tornadoes aren't the biggest storms on Earth. Still, they are among the most destructive. Scientists estimate maximum tornado winds at about 402 kilometers (250 miles) per hour. That's strong enough to pull up trees and fling cars around like toys.

Tornadoes often form at **fronts**, as very cold and very warm air masses meet. Powerful thunderstorms form. These storms can form tornadoes. The conditions for tornado formation are perfect in the middle of the U.S., especially in spring. This area is nicknamed "Tornado Alley," because it gets more tornadoes than any other place.

When forecasters see the conditions that are likely to produce a tornado, they send out a Tornado Watch. People stay tuned to radio and television reports for updates. They also watch the sky, because tornadoes can form quickly.

The average wind speed of a tornado is 150 km (96 miles) per hour. But the most powerful tornadoes can have wind speeds of more than 402 km (250 miles) per hour! That's as fast as a jet airplane.

front

A **front** is the boundary where two different air masses meet.

When a tornado is spotted, forecasters issue a Tornado Warning. Often horns or sirens will sound the alarm. People take cover right away. Family emergency plans can help people know where to go and what to do.

TECHTREK
myNGconnect.com

Digital Library

This storm cellar protects a family from deadly tornadoes.

Most families have a room in the house for shelter. The lowest floor is the safest part of the house. Some people huddle in basements. Others pick an interior part of the house away from windows or outside walls. If outside, people lie in flat, low areas away from structures.

Can scientists predict tornadoes? Not exactly. But they have tools for spotting the conditions likely to produce them. Doppler radar can detect air movement within clouds that often means the formation of tornadoes. Scientists can then warn people who might be in their path. As a result, the number of deaths from tornadoes has decreased.

SEVERE WEATHER TERMS

- **Watch:** Conditions are right for severe weather to form. Get prepared. Stay alert. Keep tuned to radio or TV for new information.
- **Warning:** Severe weather will strike your area soon. Take shelter. Evacuate if authorities say to do so.

Winter Storms Northern areas usually have cold **climates** in winter. Snow can be a common sight, but a winter storm is the most severe type of snowfall. These storms can bury roads and buildings. They create cold and icy conditions that are dangerous for people.

Winter storms are not just snowy. They bring extreme cold and high winds ,too. The most severe winter storms can last for days and bring cities to a standstill. Four days of snow in 2001 buried the city of Buffalo, in New York State, under two meters (six feet) of snow!

Snowplow trucks such as this one can spread salt and sand over road surfaces.

climate

Climate is the pattern of weather over a long period of time.

There are several types of winter **weather**. A blizzard has heavy snow and winds of at least 56 kilometers (35 miles) per hour. Blowing snow makes it hard to see ahead of you in a blizzard. Nor'easters are winter storms that form along the Atlantic. They blow in a northeast direction. In winter, they bring heavy snow, huge waves, and high winds to the coast.

Forecasters issue Winter Storm Watches when winter storms are possible in a day or two. In areas where winter storms are likely, families should keep an emergency supply kit in the house and in the car. Power failures are common. People must prepare for being stranded without heat or electricity.

A Winter Storm Warning means that the storm will begin soon or has started. Authorities tell people to stay inside. If people go out, they must dress warmly—with hats, mittens, and scarves over their mouths. This weather can be very dangerous because of the extreme cold, wind, ice, and blowing snow.

STORM WARNINGS

WINTER STORM MESSAGE	WHAT IT MEANS
WINTER STORM OUTLOOK	Winter storm conditions are possible in 2–5 days. Listen to TV or radio for storm updates.
WINTER STORM ADVISORY	Winter storm conditions may be hazardous but not severe
WINTER STORM WATCH	Severe winter storm conditions are possible in 36–48 hours. It's time to prepare.
WINTER STORM WARNING	Severe winter storm conditions have begun or will begin within 24 hours. Take steps now to protect yourself.

weather

Weather is the condition of the atmosphere at a certain place and time.

Floods Sometimes the **water cycle** produces more precipitation than usual in a region. Most inland floods happen when rain is so heavy that the ground cannot absorb it. Runoff flows over the ground into streams and rivers. As more and more runoff flows into them, their water level rises. A flood occurs when a stream or river overflows its banks onto land that is normally above water.

Floods can be more severe when the ground is already soggy. Scientists monitor rivers during periods of heavy rain. They can tell when a river is likely to flood. If they issue a Flood Watch, flooding might occur soon. People listen to TV and radio for information. They prepare to leave low areas on the coast or near rivers.

If a Flood Warning is issued, flooding will happen soon. Authorities tell some people to evacuate. If a Flash Flood Warning is issued, people must get to higher ground right away—by car or even on foot. A wall of water can rush down a stream quickly in a flash flood.

A powerful winter storm and record amounts of rain caused rivers and creeks to overflow in Napa, California.

BE AWARE

- Floods can occur in any season
- Cars and pickup trucks can be carried away by just half a meter (two feet) of moving floodwater
- Flash floods can happen in areas where there has been no rain

water cycle

The **water cycle** is the constant movement of Earth's water from the surface to the atmosphere and back again.

Many people in areas that flood build the first floor of their houses high off the ground. But the best way to prepare for floods is to avoid building in low areas near streams or on beaches. In some places, people no longer put homes, schools, or stores in these places. They are left for parks and land uses that flooding will not damage.

Weather forecasters warn the public about many types of severe weather. As with other severe weather situations, it is best to be prepared. Families should have their own emergency plan. These plans tell everyone what to do in a weather emergency. Families should also have an emergency supply kit.

Scientists use stream-flow gauges to monitor the water level in streams.

People place sandbags along streams to hold back floodwaters.

NATIONAL GEOGRAPHIC

SHARE AND COMPARE

Turn and Talk Are certain types of severe weather more common in some climates than in others? Form a complete answer to this question together with a partner.

Read Select two pages in this section that are the most interesting to you. Practice reading the pages so you can read them smoothly. Then read them aloud to a partner or small group. Talk about why the pages are interesting.

my SCIENCE notebook **Write** Write a conclusion that tells the important ideas about what you have learned about severe weather. State what you think is the Big Idea of this section. Share what you wrote with a classmate. Compare your conclusions. Did your classmate recall that hurricanes only form over oceans?

my SCIENCE notebook **Draw** Imagine what it would be like to experience a hurricane, flood, blizzard, or tornado. Draw a picture of an activity that you could do to prepare for it. Combine your drawing with those of your classmates to make your own emergency plan.

FLORIDA PHYSICAL SCIENCE

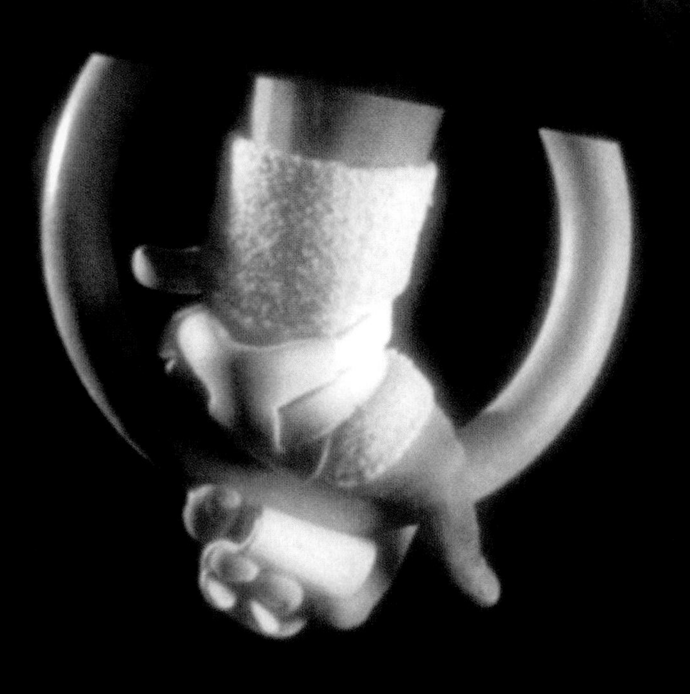

What Is Physical Science?

Physical science is the study of the physical world around you. This type of science investigates the properties of different objects, as well as how those objects interact with each other. Physical science includes the study of matter, motion and forces, and many kinds of energy, including light and electricity. People who study how all of these things work together are called physical scientists.

You will learn about these aspects of physical science in this unit:

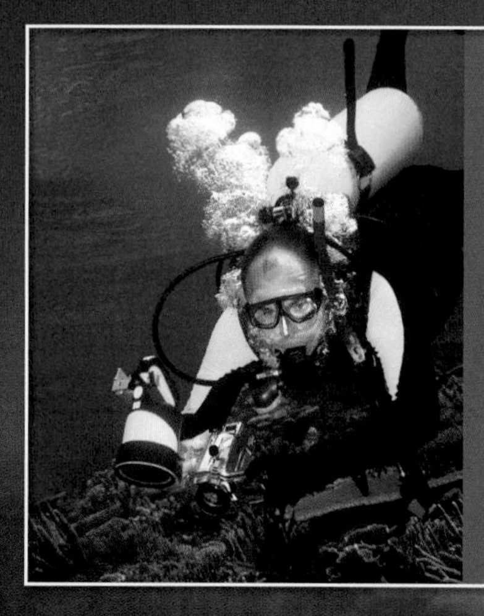

HOW CAN YOU DESCRIBE MATTER, MIXTURES, AND SOLUTIONS?

Matter is anything that has mass and takes up space. Physical scientists study all of the different properties of matter. These include size, shape, color, texture, hardness, and temperature, as well as mass and volume. Physical scientists also study how matter can undergo both physical and chemical changes. Different mixtures and solutions are often the results of these changes.

HOW DO YOU DESCRIBE FORCE AND THE LAWS OF MOTION?

Physical scientists study how objects move. They also study the forces that act on objects, and how those objects respond to different kinds of forces. Isaac Newton was a famous physical scientist who described his observations of the effects of force on objects into what is now known as the laws of motion.

HOW DO YOU DESCRIBE DIFFERENT FORMS OF ENERGY?

Physical scientists study energy in all of its forms. They also learn about how energy can cause changes in the physical world. The different kinds of energy include light, sound, electrical, and mechanical.

HOW DOES ELECTRICAL ENERGY FLOW AND TRANSFORM?

Many objects in our world depend on electrical energy for the energy needed to function. Physical scientists study which objects electricity flows through easily, and which objects it does not. They also study how electricity can change into other forms of energy, such as light and sound.

MEET A SCIENTIST

Thomas Taha Rassam Culhane: Urban Planner

T.H., as he likes to be called, is an urban planner and National Geographic Emerging Explorer. T.H. and his team work in some of the poorest neighborhoods of Cairo, Egypt, and elsewhere to build and install rooftop solar water heaters, kitchen-waste-to-biogas digesters and other devices to meet people's everyday needs for cooking fuel, electricity, and clean water.

T.H. explains. "The technology we create and install with the people is completely carbon dioxide free, it contributes nothing to global warming. If people don't have access to enough water, it becomes a serious health issue. And when women spend all their time tending stoves to heat water, how can they go to school or get ahead?" T.H.'s grasp of these daily challenges is personal; he and his wife once moved to a slum apartment themselves to gain firsthand experience.

T.H. uses kitchen scraps and waste water that would have normally been put in the trash or down the drain to create cooking fuel in a biogas digester. The digester provides enough fuel for a family to cook their meals.

In Chapter 5, you will learn:

FLORIDA NEXT GENERATION SUNSHINE STATE STANDARDS

SC.5.P.8.1 Compare and contrast the basic properties of solids, liquids, and gases, such as mass, volume, color, texture, and temperature. **PROPERTIES OF MATTER, MASS AND VOLUME**

SC.5.P.8.2 Investigate and identify materials that will dissolve in water and those that will not and identify the conditions that will speed up or slow down the dissolving process. **SOLUTIONS**

SC.5.P.8.3 Demonstrate and explain that mixtures of solids can be separated based on observable properties of their parts such as particle size, shape, color, and magnetic attraction. **MIXTURES**

SC.5.P.8.4 Explore the scientific theory of atoms (also called atomic theory) by recognizing that all matter is composed of parts that are too small to be seen without magnification. **ATOMS**

SC.5.P.9.1 Investigate and describe that many physical and chemical changes are affected by temperature. **PHYSICAL AND CHEMICAL CHANGES**

SC.5.P.8.4 Science in a Snap! Explore the scientific theory of atoms (also called atomic theory) by recognizing that all matter is composed of parts that are too small to be seen without magnification.

CHAPTER

5

HOW CAN YOU DESCRIBE MATTER, MIX

Look around you. What do you see? You see matter. Matter is everywhere. You can see it in the forms of liquids, solids, and gases. What matter can you observe in this picture?

Solids and liquids are easy to see. The diver's scuba gear is a solid. The ocean water is a liquid. But what about gases? The bubbles you can see around the diver are gas.

TECHTREK
myNGconnect.com

Student eEdition
Vocabulary Games
Digital Library
Enrichment Activities

SCIENCE VOCABULARY

mass (MAS)

Mass is the amount of matter in an object. (p. 206)

The mass of the ball on the left is the same as the mass of the gram weights on the right.

volume (VOL-yum)

Volume is the amount of space something takes up. (p. 208)

A beaker is a tool that measures the volume of a liquid.

atom (A-tum)

An **atom** is the smallest piece of matter that can still be identified as that matter. (p. 210)

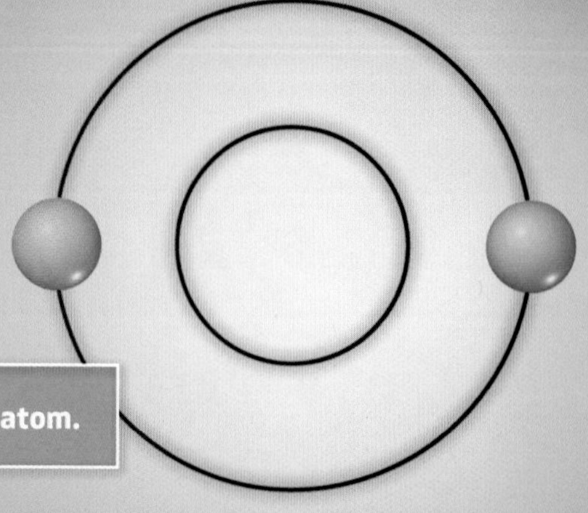

This is a model of an atom.

my Science Vocabulary

atom (A-tum)

mass (MAS)

mixture (MIKS-chur)

solution (so-LŪ-shun)

volume (VOL-yum)

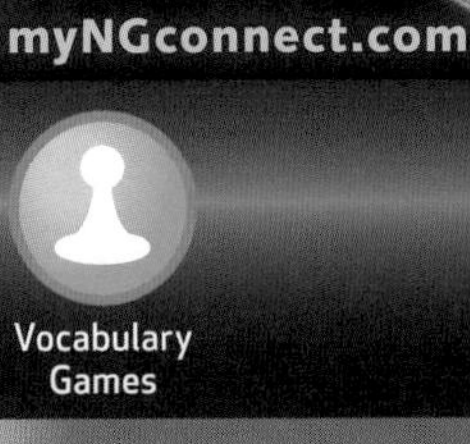

mixture (MIKS-chur)

A **mixture** is two or more kinds of matter put together. (p. 212)

This mixture of nuts and bolts can easily be sorted by color and size.

solution (so-LŪ-shun)

A **solution** is a mixture of two or more kinds of matter evenly spread out. (p. 216)

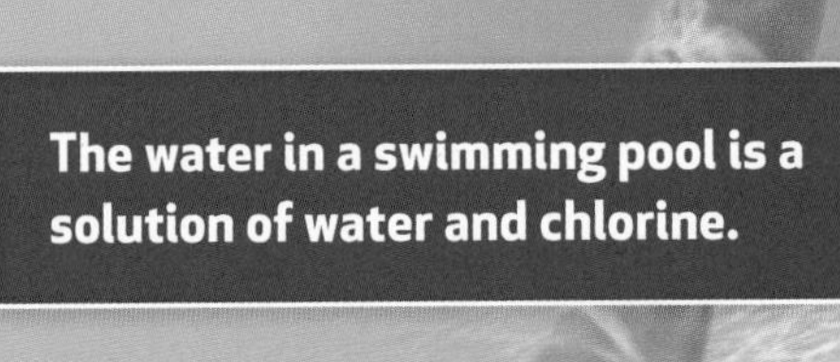

The water in a swimming pool is a solution of water and chlorine.

Properties of Matter

How do scientists define *matter*? Matter is anything that has mass and takes up space. Scientists describe and compare matter based on properties, or qualities. State, color, texture, temperature, mass, and volume are some of matter's properties.

State Solids, liquids, and gases are called states of matter. While ice is a solid, for example, water is a liquid.

Some matter, like a block of wood, has a definite shape and volume. A piece of wood is a solid. The tightly-packed particles in all solids give them their shapes.

If the particles in the matter are able to move around, the matter may be liquid. Liquids like water, juice, and milk have definite volumes and take the shapes of their containers.

If you've ever seen a balloon floating in the air, you've seen an object filled with gas. Gases have no definite shape or volume. Instead, their quickly-moving particles spread out to fill a space.

Particles in solids are packed tightly together.

Particles in liquids are able to move, though they bump into each other.

Particles in gases are spread far apart, move rapidly, and sometimes bump into each other.

Air is a gas. Moving air can push objects, like this sail.

The ocean is a liquid.

The surfboard is a solid.

Color and Texture Color and texture are two properties of matter that scientists describe and compare. Solids, liquids, and gases may all have color. Solids and liquids can have texture.

Light allows you to observe the property of color. *Red, blue,* and *yellow* are just a few words that describe color.

If you've ever run your hand over sandpaper or fur, you've felt texture. Textures, like rough, smooth, bumpy, and furry, are properties you can feel.

Flowers come in a variety of colors.

The texture of a tennis ball can be described as soft, furry, or fuzzy. A baseball's texture is smooth and leathery. Basketballs have a bumpy or rough texture.

Temperature Your tea is too hot, so you add ice to make it cooler. You are changing the temperature of your drink. Temperature is a measure of how hot or cold an object is. How can scientists measure and describe the property of temperature? They use thermometers and two different scales. The Fahrenheit scale is the scale used most often in the United States. People in other countries, and scientists around the world, measure temperature using the Celsius scale.

Water freezes to form ice at 0°C (32°F).

Iron ore is heated to 871°C (1,600°F) to create steel.

Before You Move On

1. What properties do scientists use to describe and compare matter?
2. How would you describe the color, texture, and temperature of your science book to a friend?
3. **Analyze** How could properties such as color, texture, and temperature help scientists distinguish between two different samples of matter?

Mass and Volume

Mass Some big objects, such as trucks, have a lot of matter. Smaller objects, such as toy trucks, have much less matter. How can you measure the amount of matter in an object? You can figure out its **mass**. Mass is the measure of the amount of matter in an object. Mass is measured using metric units: grams and kilograms. To find the mass of an object, compare it to grams and kilograms on a tool such as a pan balance. Place the object on one side of the balance. Add gram weights to the other side until the sides are balanced. Add the sum of the gram weights to find the mass of the object.

Look at the picture below. What do you notice about the two sides of the pan balance? What does this tell you about mass?

TECHTREK
myNGconnect.com

Digital Library

You can use a pan balance to measure the mass of an object.

Mass is different from weight. Your weight is a measure of gravity's pull on you. On the moon you would weigh less than on Earth because the moon's gravitational pull on your body is less. Your weight on the moon would be one-sixth of your weight on Earth. On Jupiter, which has a larger gravitational pull, you would weigh more, about 2.5 times more than you weigh on Earth. But you would still have the same mass in all three places. The only thing that has changed is gravity's pull.

The smaller force of gravity on the moon would affect your weight. However, your mass would remain the same.

The force of gravity on Earth affects your weight, but not your mass.

Volume Matter takes up space. The amount of space taken up by an object is its **volume**. You can measure the volume of solids, liquids, and gases.

Calculating Volume If you want to find the volume of an object like a block, you can measure it and then use math. Measure the length, width, and height and then multiply the three numbers.

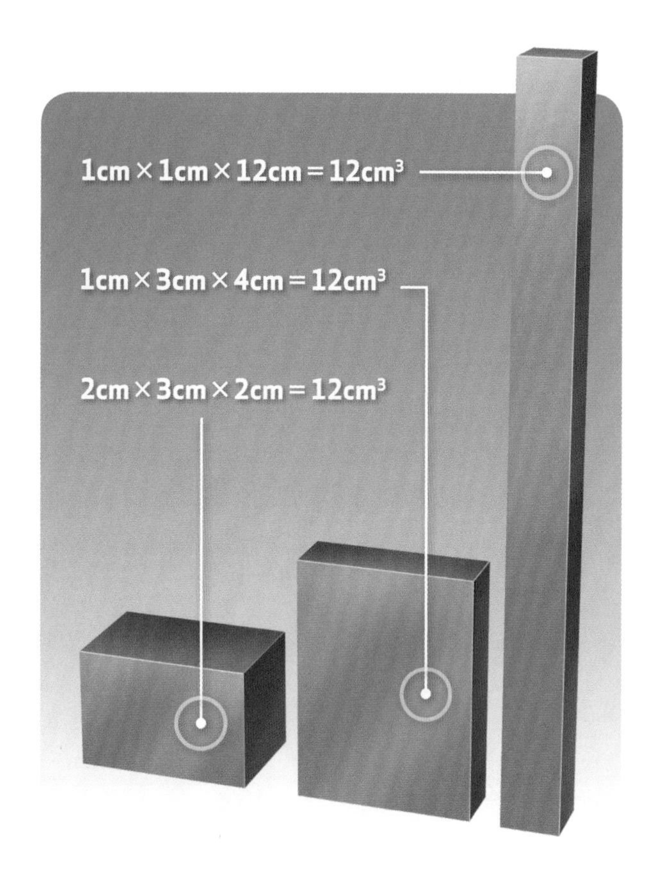

How would you measure the volume of a marble or rock? Fill a beaker of water and note the volume. Then, place the object in the beaker. Note the new reading. To find the volume of the solid object, find the difference between the two volumes.

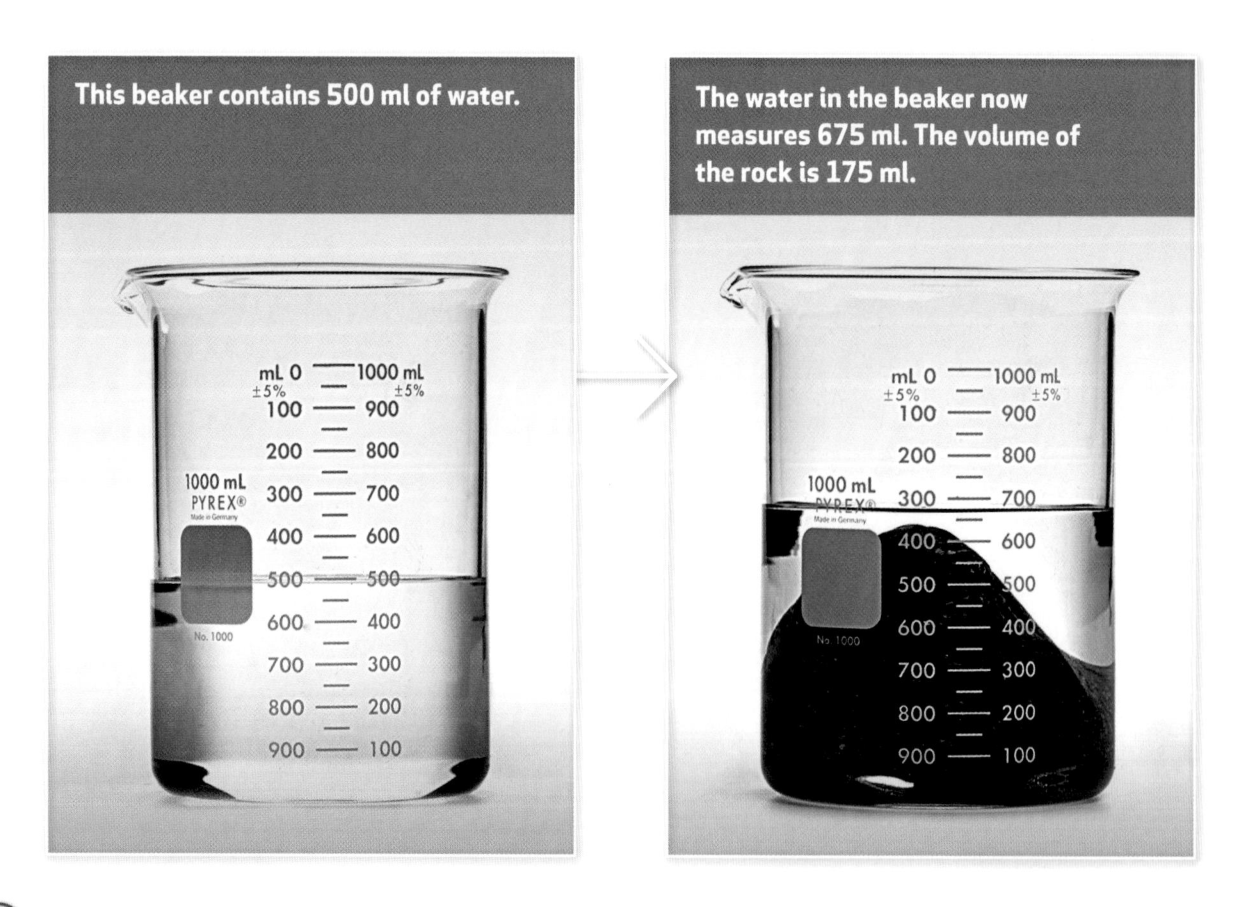

If you were thirsty, would you want to drink 1 milliliter (mL) or 1 liter (L) of water? Both of these units measure volume. Milliliters are very small!

Liquids have a definite volume, but they take the shape of their containers. Look at the picture. The volume of the liquid is the same in both containers, but the containers are different sizes.

Volume of Gases Gases spread out to fill their containers. The volume of gas is the same as the size of its container. The volume of air in a liter bottle, for example, would be one liter.

Enrichment Activities

Because liquid takes the shape of its containers, the volume of liquid looks different in each container.

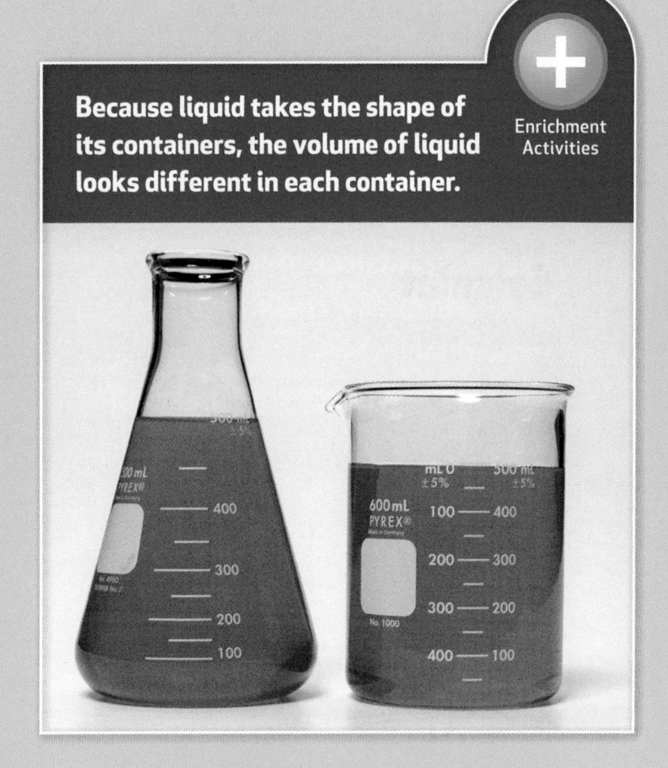

One propane container has the same volume as nine two-liter bottles.

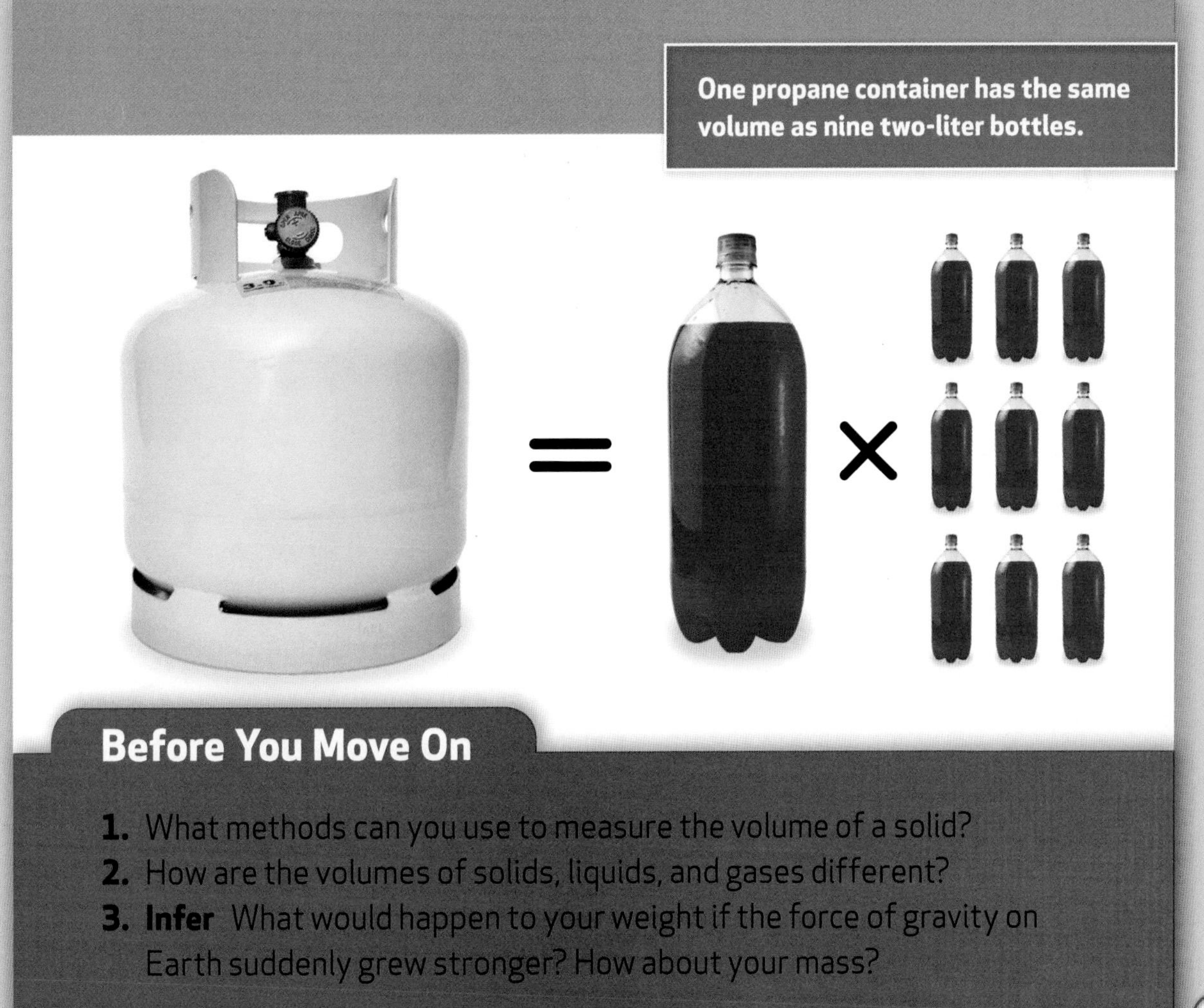

Before You Move On

1. What methods can you use to measure the volume of a solid?
2. How are the volumes of solids, liquids, and gases different?
3. **Infer** What would happen to your weight if the force of gravity on Earth suddenly grew stronger? How about your mass?

Atoms

Think about your favorite book. What is the smallest part that makes up the book? Writers use small pieces—letters—to create words, sentences, paragraphs, and, eventually, the story. Like individual letters, **atoms** are the smallest parts that make up matter. Everything in our world—you, this book, Earth itself—is made up of tiny atoms.

There are over 100 different kinds of atoms in the universe. Every atom has three parts: protons, electrons, and neutrons. Atoms are too small to be seen.

This balloon will soon be full of atoms of gas!

The protons in an atom are in the nucleus. A neutral atom has the same number of electrons as protons.

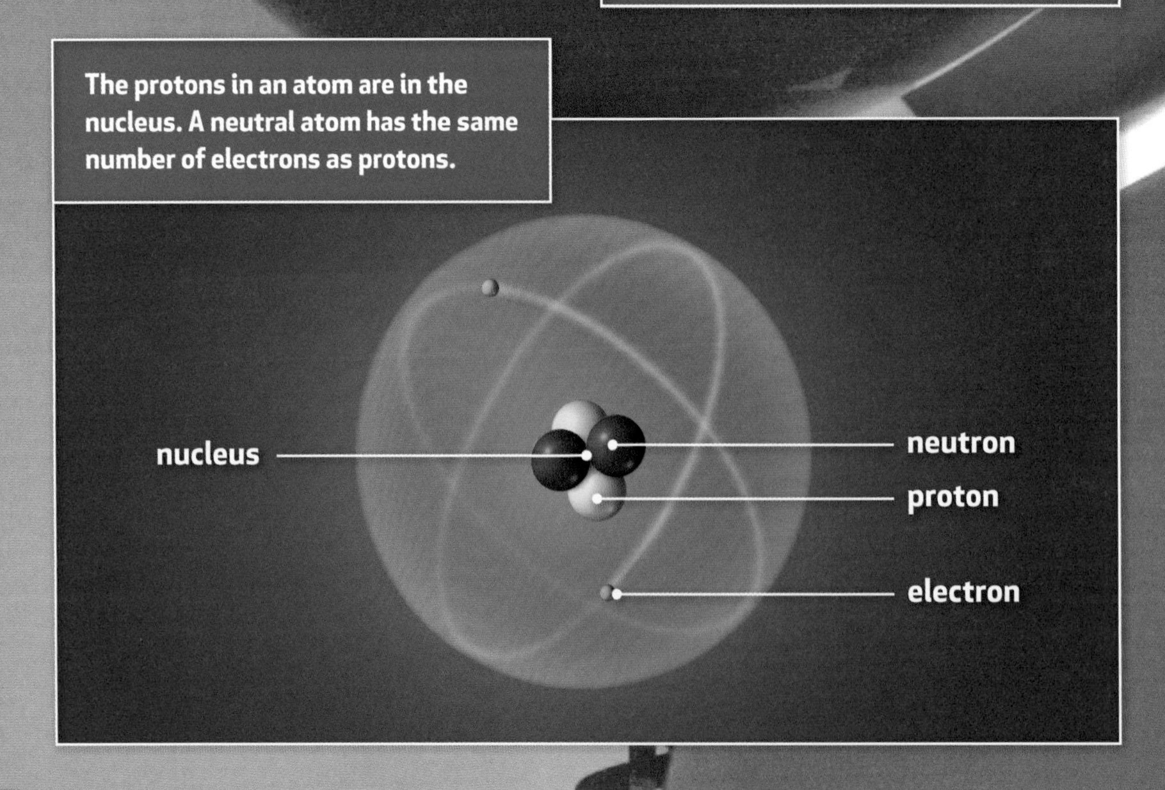

Protons exist in the center, or nucleus, of the atom. Protons have a positive electrical charge. The number of protons in the nucleus identifies the atom. For example, helium atoms have two protons.

Neutrons exist with protons in the nucleus of the atom. Neutrons have no electrical charge.

Electrons move around the nucleus of an atom. Electrons have a negative electrical charge. In a neutral atom, the number of electrons in an atom is equal to the number of protons in the nucleus.

Science in a Snap! Close Up

Use a hand lens to closely examine fabric, such as on your shirt or sweater.

Describe what you see.

How did your view of the fabric change with the hand lens?

Before You Move On

1. What are the parts of an atom?
2. How are atoms related to matter?
3. **Compare and Contrast** How are protons and neutrons alike? How are they different?

Mixtures

Look at the nuts and bolts on this page. They are not all alike. They are a **mixture**, two or more things that are combined. A mixture can be separated by its properties, such as size, color, and shape. What properties would you use to separate the mixture of nuts and bolts?

Mixtures are made up of different amounts of two or more substances. There do not have to be equal amounts, and the amount of each different substance can change. Each part of a mixture keeps its original properties. The parts of a mixture can be easily separated.

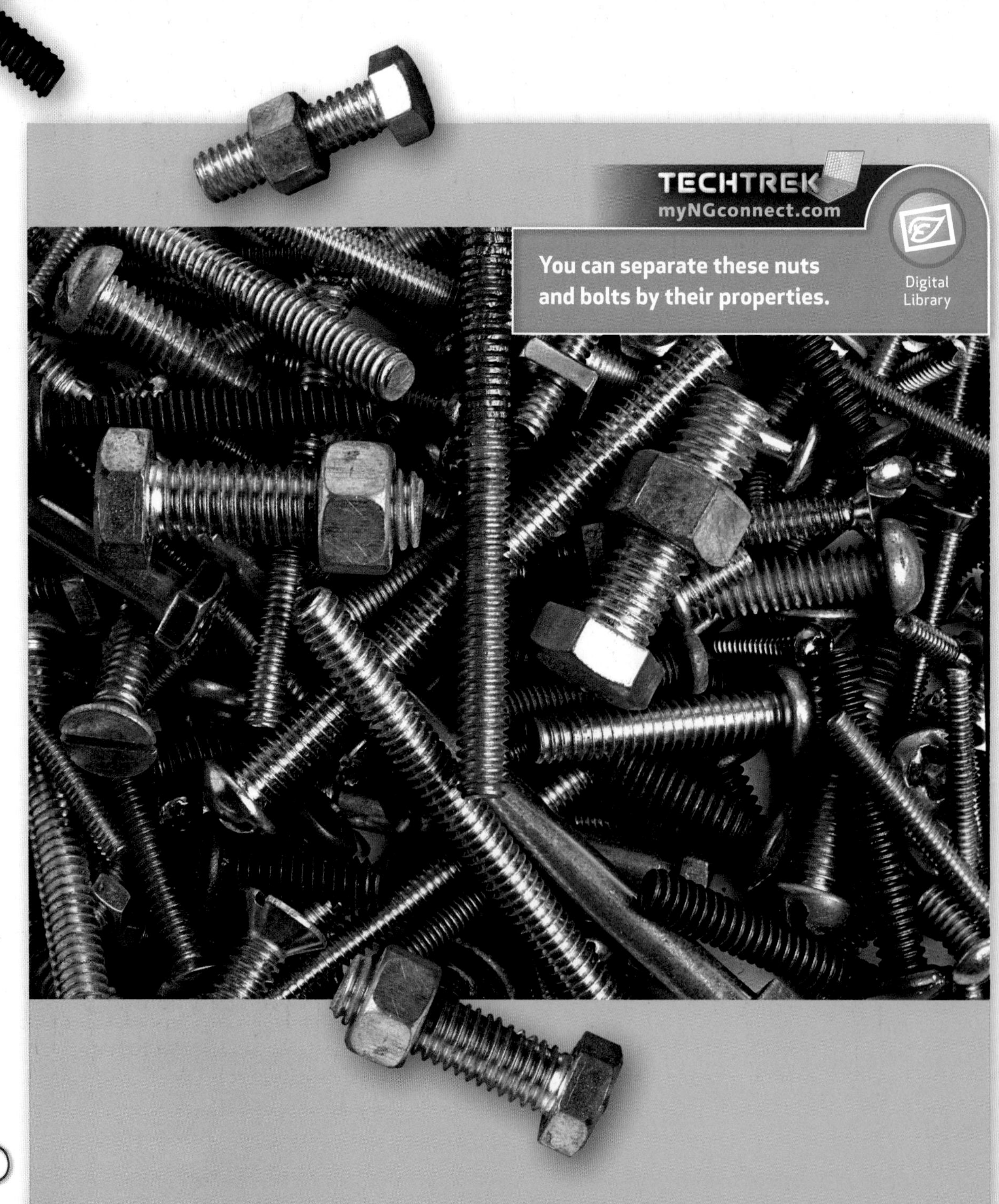

You can separate these nuts and bolts by their properties.

You can separate the mixture of nuts and bolts by color and size. There are many ways to separate mixtures according to their properties.

Look at the mixture on this page. Some of the objects are magnetic, and others are not. You could use a magnet to separate some parts of the mixture. The paper clips, pins, and tacks would separate from the rubber bands and plastic figures. Once the magnetic items are sorted from the ones that do not attract magnets, how could you separate items in each group based on their properties?

The magnets attracts all of the magnetic objects. The non-magnetic objects do not contain iron. They are left on the table.

Types of Mixtures

- Look at the picture of the mud. See the liquid and the solid particles sticking to the girl's hands? Mud is a **solid/liquid** mixture. Cake batter and lemonade are other mixtures of liquids and solids.
- A **liquid/liquid mixture** is a combination of two liquid ingredients. Oil and vinegar salad dressing is a mixture that contains only liquids. So is gasoline.

Oil and vinegar salad dressing is a liquid/liquid mixture. The oil is on top of the vinegar in the photograph.

- Liquid matter and gases can combine to form **liquid/gas** mixtures. Soda water and bubbles found in foaming baths and dishwater are examples of mixtures that contain both liquids and gasses.
- The cereal is a mixture of flakes, nuts, and dried fruits. **Solid/solid** mixtures contain all solid parts. Soil is another mixture made only of solids.

Bubbles containing gas form in this dishwater.

Before You Move On

1. What are the properties of a mixture?
2. You are holding a glass of orange juice with ice cubes floating in it. What kind of mixture are you holding?
3. **Analyze** If you found an interesting mixture, what properties would you look for to try and separate it into its individual components?

Solutions

If you have ever swum in a pool, you have swum in a solution! A **solution** is a mixture of two or more types of matter evenly spread out and not easily separated. Every sample of the solution will contain the exact same parts, properties, and appearance. A solution is made up of two parts: a solute and a solvent.

In a swimming pool, liquid chlorine is added to water to make the water safe and clean for swimming. The water is the solvent—there is more water than liquid chlorine. The liquid chlorine is dissolved in the water to make a solution. That makes the chlorine the solute.

Pool water is an example of a solution made up of two liquids: water and liquid chlorine.

Solutions can also contain materials from different states. Juice made from a powder mix, for example, is a solution of a liquid solvent (water) and a solid solute (juice mix). Solubility is matter's ability to dissolve in a liquid. Look at the images of solids below. Which do you think are soluble in water? Which are not?

Factors that affect how fast a solid dissolves include:

- Size: smaller solids dissolve faster.
- Temperature: solids dissolve faster at higher temperatures.
- Stirring: solids dissolve faster when the solution is stirred.

When salt or sugar is added to water, it dissolves and evenly distributes throughout the water creating a solution. When you look at the solution, you cannot see the particles of salt or sugar. When you add sand or a coin to water, you have created a mixture. The sand and coin do not dissolve and distribute through the water. At some points in the mixture, you will find just water. At other points, you will find sand or a coin.

Concentration and Saturation Look at the three glasses of juice on the next page. They are all made with water and juice mix. What do you notice about them? They are different colors, because each of them has a different amount of juice mix. The juice glasses have solutions with different concentrations. The concentration of a solution is the amount of solute that is dissolved in the solution.

The juice on the left has a low concentration. It has a small amount of juice mix compared to the amount of water in the glass. The second picture shows a medium concentration, while the third picture shows a high concentration of juice mix.

Scientists can classify solutions based on their concentrations. A concentrated solution has a lot of solute in it. The juice on the far right is the most concentrated solution.

The water in this Florida swamp contains a high concentration of dirt and other materials.

Dilute means "thinned." The middle glass of juice could be described as a dilute solution. The glass of juice on the left, the weakest solution, would be described as very dilute.

Sometimes, solute is added to a solvent until the solvent cannot dissolve any more solute. If you kept adding juice mix to the water, for example, eventually the water would stop dissolving the mix. Extra juice mix would stay solid, sinking to the bottom or floating to the top of the liquid.

Before You Move On

1. What are the parts of a solution?
2. How does the amount of a solute in a solvent affect the solution's concentration?
3. **Infer** What do you think happens when Earth's atmosphere becomes saturated with water vapor?

Physical and Chemical Changes

You can find water on Earth in three different states: as a liquid, as a solid (ice), and as a gas (water vapor). No matter what form it's in, water is made up hydrogen and oxygen atoms.

Water changes when the temperature changes. Theses changes are called physical changes. The water looks different, but the atoms that make up the water do not change. Physical changes can be reversed. When water freezes, for example, it becomes ice. But when the temperature goes up, the ice changes back to a liquid.

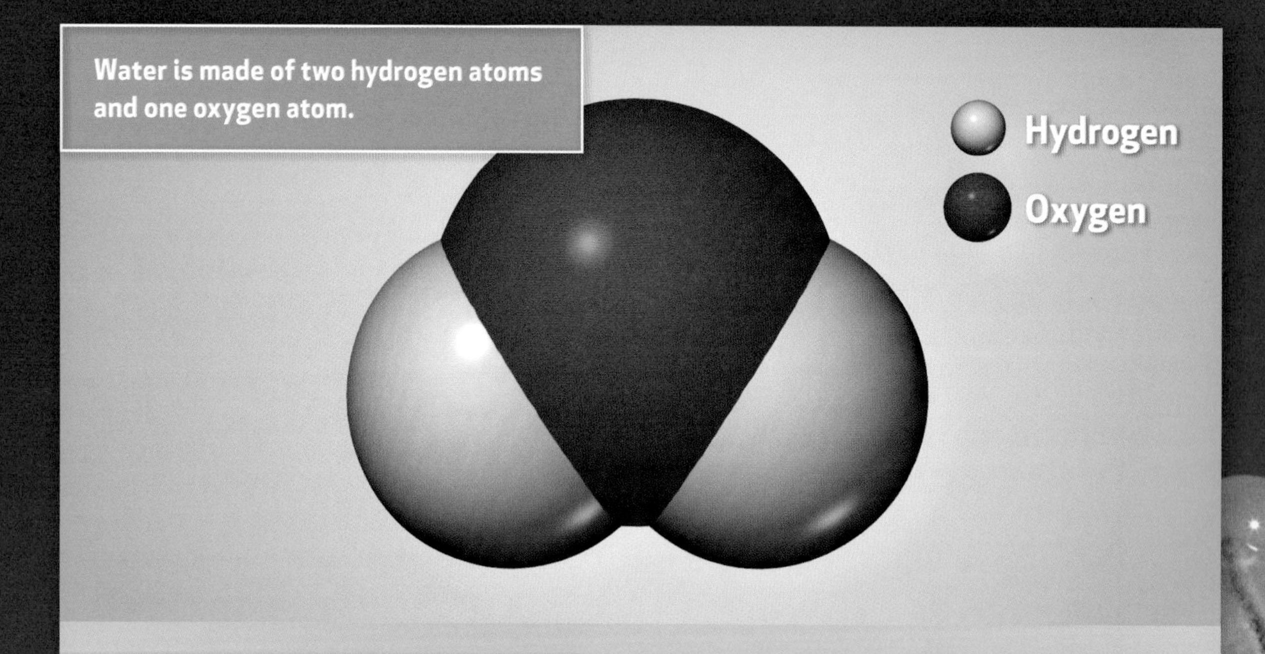

Water is made of two hydrogen atoms and one oxygen atom.

When water boils, the molecules move faster and faster. Some molecules break away and form water vapor.

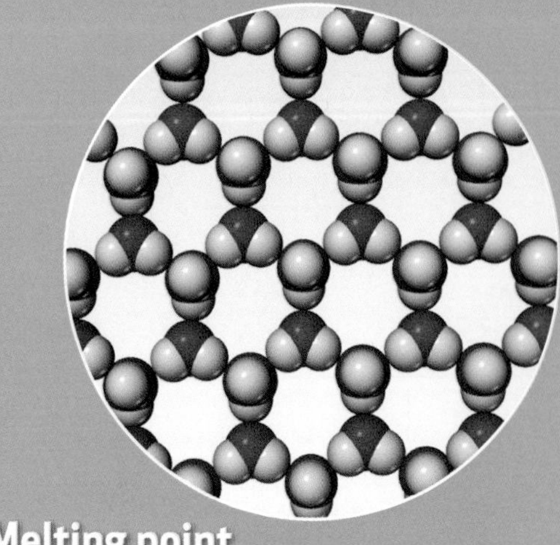

Melting point

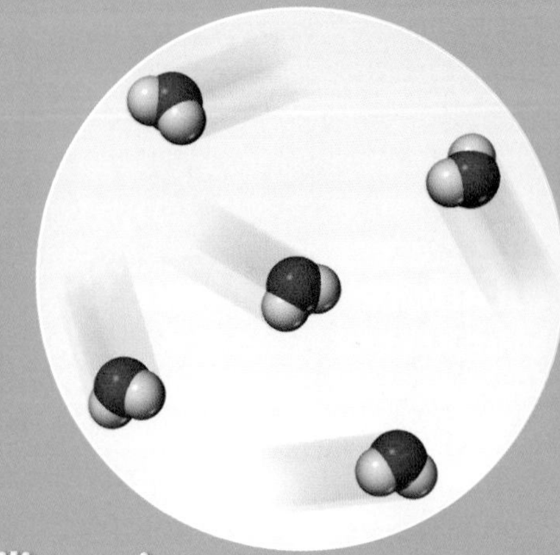

Boiling point

Freezing and Melting Water is a liquid at room temperature. Think of a river or a stream. Liquid water flows. When the temperature drops to 0°Celsius (32° Fahrenheit), the water starts to freeze. It no longer flows. It becomes solid. This temperature is water's freezing point. As you freeze a liquid, the atoms move at a slower rate. The liquid becomes a solid. Zero degrees Celsius (32°F) is also the melting point of ice. When the temperature of frozen water rises to 0°C, the atoms begin to move faster. The water returns to the liquid state.

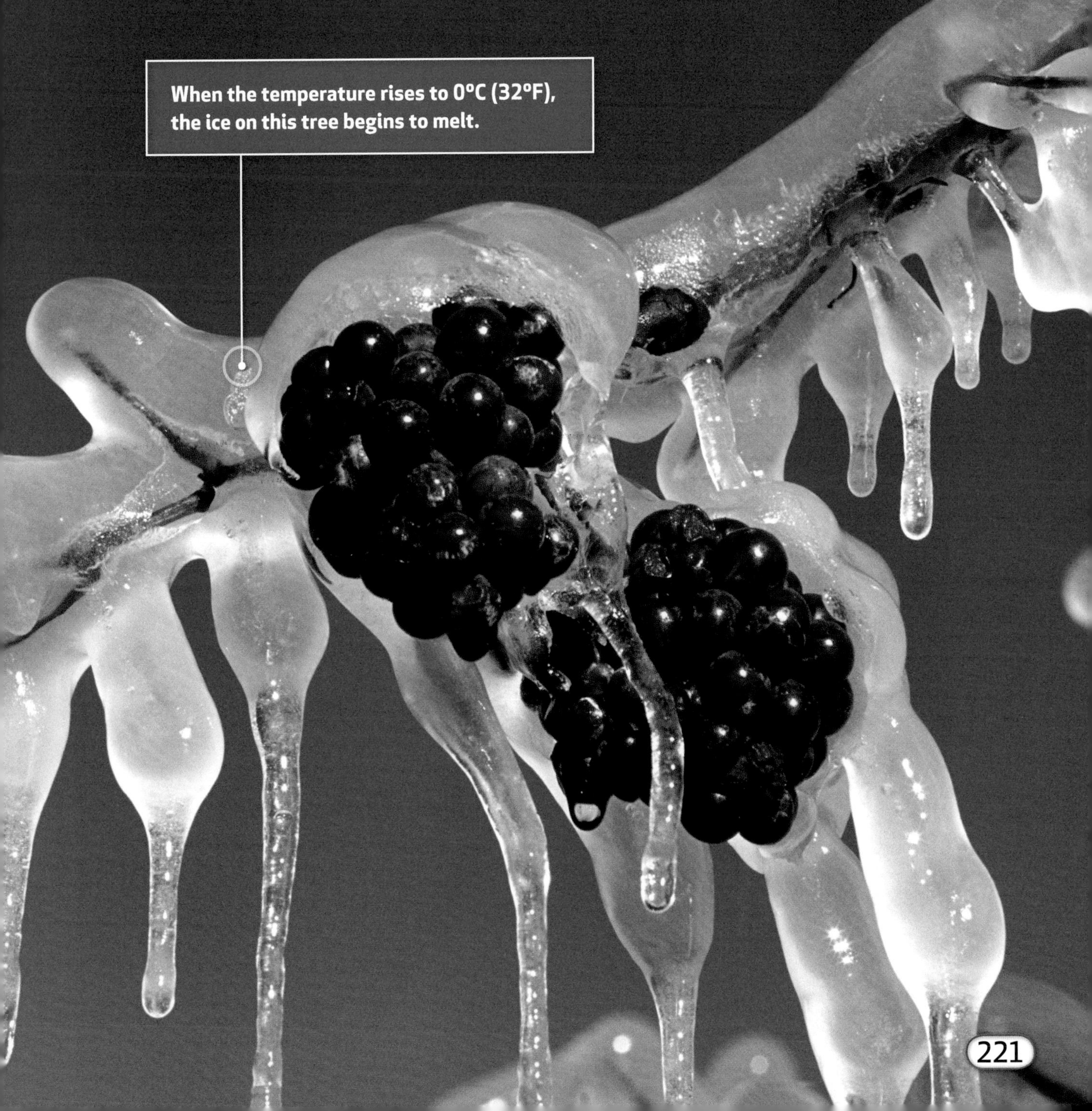

When the temperature rises to 0°C (32°F), the ice on this tree begins to melt.

Evaporation Have you noticed that when water boils, you can see bubbles? When heat is added to water in the liquid state, it causes water molecules to move faster and faster. The liquid water changes into a gas. This process is called evaporation. Evaporation is the physical change of matter from a liquid state to a gaseous state.

Evaporation occurs when the temperature of a liquid is raised enough to change the liquid to a gas. Water boils at 100°C (212°F). This is called the boiling point.

When water is heated to its boiling point, the liquid water changes to a gas. We call this water vapor.

Water vapor condensed to form dew droplets on this spider web.

Condensation Condensation is the opposite of evaporation. It is the physical change of matter from a gaseous state to a liquid state. You can see examples of condensation in clouds, fog, and dew on the grass. Have you ever wondered why your glass of ice water "sweats"? The warm water vapor in the air touches the cold glass. The cold of the glass causes the temperature of the water vapor to drop. The water molecules slow down, and the vapor changes from a gas to a liquid on the side of your glass. The same thing happens when warm daytime air is cooled at night. The condensation you see and feel on the grass is called dew.

Condensation is formed by water vapor in the air touching the cold sides of the glass.

Chemical Changes In addition to physical changes, matter can undergo chemical changes. A chemical change is a change in matter that forms a new substance with different properties. When a chemical change takes place, the matter has been changed permanently. Matter may make one or more of the changes listed the table.

SIGNS OF A CHEMICAL CHANGE

- √ gas forms
- √ heat is given off or absorbed
- √ a solid forms or disappears
- √ light is produced

When wood burns, light and heat are produced.

Burning Burning wood requires oxygen and heat. If you burn wood, the wood is fuel. What happens when wood is burned? First, the wood must be heated to a very high temperature. The heat makes the wood change into a different type of matter.

Burning causes a chemical change in the wood. How do we know this? Look at the signs listed in the chart. Gas forms (smoke), heat is given off, a solid (wood) disappears and a new solid (ash) forms, and light is produced. These changes are permanent. No one can ever change ash back into wood.

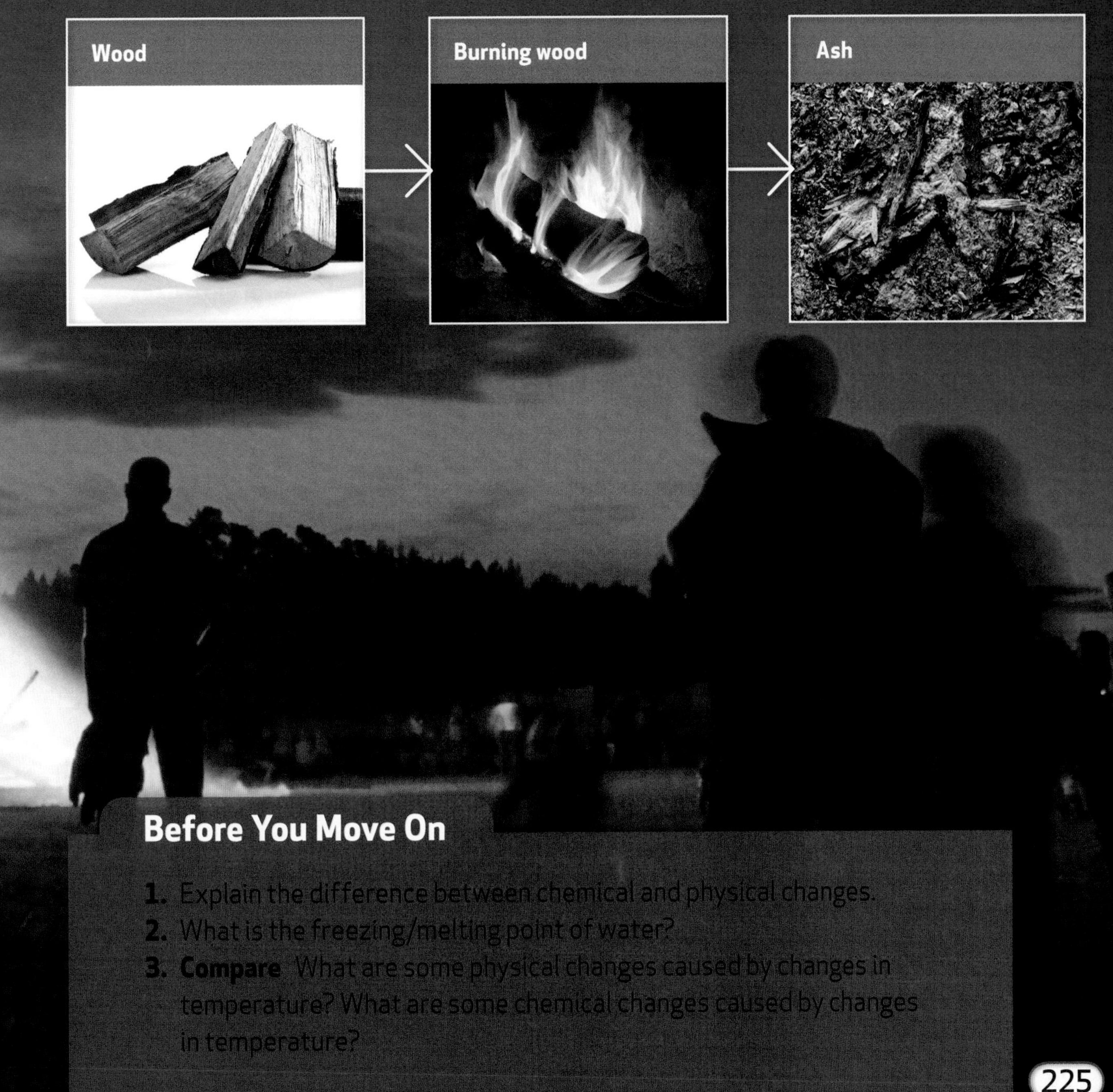

Before You Move On

1. Explain the difference between chemical and physical changes.
2. What is the freezing/melting point of water?
3. **Compare** What are some physical changes caused by changes in temperature? What are some chemical changes caused by changes in temperature?

NATIONAL GEOGRAPHIC

A NATURAL SOLUTION

The Dead Sea is one of the smallest seas in the world, less than 80.5 km (50 miles) long ranging from 3.2 km (2 miles) to 17.7 km (11 miles) wide. Located between Jordan and Israel in southwestern Asia, the Dead Sea is surrounded by land. The name "Dead Sea" is a gentle translation from its Hebrew name, "Yam ha Maved," which literally means "Killer Sea." At 672.7 km (418 miles) below sea level, the Dead Sea has the lowest water surface in the world.

The Dead Sea is the saltiest body of water in the world. The Jordan River and some smaller streams feed the surface of the Dead Sea. The fresh water from these sources is quickly evaporated in the hot desert climate. At the surface, the water the least salty. As the water gets deeper, it gets much saltier. At about 40 meters (130 feet) deep, the Dead Sea is about ten times as salty as the oceans. Deeper, below 90 m (300 feet) the sea is completely saturated. The water at this point cannot hold any more salt. The salt settles in drifts at the bottom of the sea.

The Dead Sea is in Israel. The concentration of salt is so high that people float on the water.

Salt collects along the shore of the Dead Sea.

No rivers or streams lead out of the Dead Sea to the sea. With no way for the water to escape, it just becomes saltier and saltier. How? The fresh water that feeds into it settles on the surface and quickly evaporates, leaving behind minerals. These minerals sink into the water below, increasing the concentration of salt in the water.

No animals live in or around the Dead Sea. The environment is too harsh. A few bacteria and algae have adapted to its severe conditions. Fish that accidentally swim into the sea from the Jordan River or other streams die instantly. Their bodies become covered in salt crystals and they wash up on shore. The fish become part of the white, salt-covered landscape that surrounds the sea.

The concentration of salt in the Dead Sea is so high that the water is much denser than ocean or fresh water. This allows humans to simply float at the surface like corks. Because of the dense water, it is actually difficult to swim in the sea. Most people who visit simply lie back in the water, float around, and read. People with arthritis and skin problems also visit the Dead Sea. The seawater helps them feel better.

Conclusion

Matter is anything that has mass, takes up space, and occurs as a solid, liquid, or gas. Scientists look at the properties of matter when describing or comparing. These properties include color, texture, temperature, mass, and volume. Matter is made up of particles called atoms. Matter can combine to form mixtures and solutions. Matter can also change. Physical changes are caused by temperature. Chemical changes create new kinds of matter.

Big Idea Everything in the universe is made up of matter, which is made up of atoms.

The helium inside this balloon is matter in the gas state.

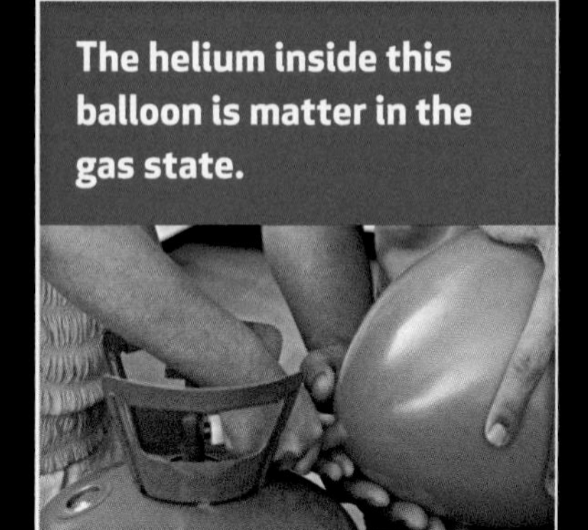

Water is matter in the liquid state.

Bricks are matter in the solid state.

Vocabulary Review

Match the following terms with the correct definition.

A. Mass
B. Volume
C. Atom
D. Mixture
E. Solution

1. The amount of space something takes up
2. The smallest piece of matter that can still be identified as that matter
3. Two or more kinds of matter put together
4. The amount of matter in an object
5. A mixture of two or more kinds of matter evenly spread out

Big Idea Review

1. **List** List at least three properties you can use to describe matter.

2. **Describe** How is weight different from mass?

3. **Compare and Contrast** How is a solution of salt and water different from a mixture of salt and gravel?

4. **Cause and Effect** You and a friend are making a pitcher of lemonade. The directions call for 1 liter of water, but you add 2 liters by mistake. How will that affect the properties of the solution?

5. **Analyze** If you had a mixture of marbles, how could you separate them into groups based on their properties?

6. **Generalize** You have an unknown solid substance. It is shaped so that it cannot be measured with a ruler. What properties can you use to describe the object? How can you measure it?

Write About the Properties of Matter, Mixtures, and Solutions

Explain Is this a mixture or a solution? How do you know? Explain how you could separate these items.

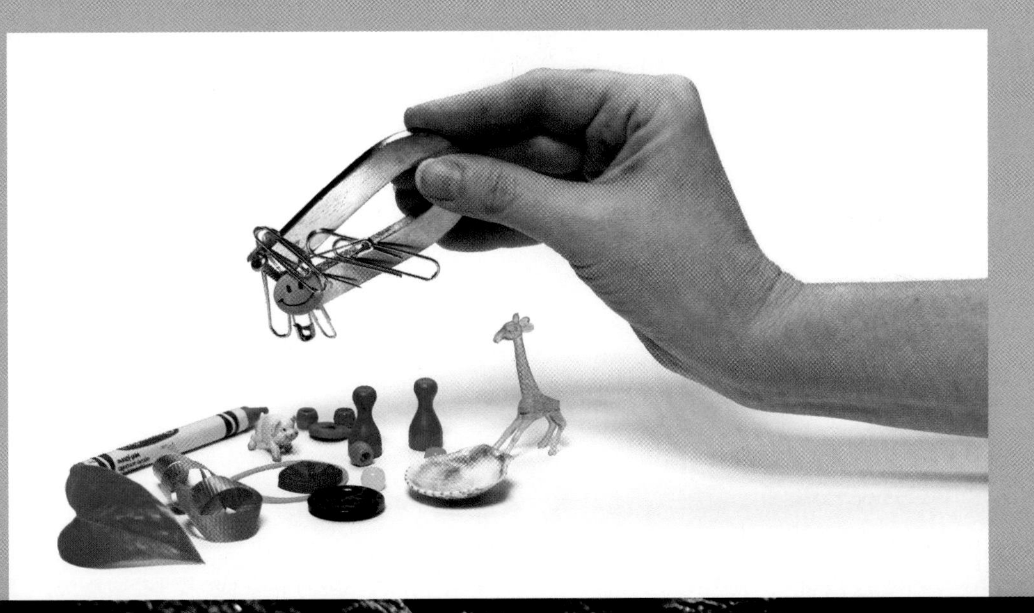

PHYSICAL SCIENCE EXPERT: CHEMIST

Have you ever thought about the way a plant makes food? It takes energy from the sun to make the food. The food is a new kind of matter that it needs to live and grow.

Tehshik Yoon is using this same idea to make new kinds of matter. He is a teacher at a university. Dr. Yoon and his research group ask a lot of questions. They want to know the best way to capture sunlight. They are also trying to find out how to use sunlight to make new kinds of matter.

Dr. Yoon teaches students about matter.

TECHTREK
myNGconnect.com
Student eEdition
Digital Library

TECHTREK
myNGconnect.com
Digital Library

Dr. Yoon uses a special machine to help him make new kinds of matter.

Dr. Yoon has always been interested in building things. As a child, he spent his time building structures out of blocks. Now he builds structures that are so small you cannot see them!

Dr. Yoon's work is very important in many fields, but it is especially important in making medicines. Now people will be able to make medicines using sunlight which is a clean source of energy. Many chemical plants have to burn oil and use electricity in order to make medicines. This can be harmful to the environment. Using sunlight to make the same medicines cost less money and causes less pollution. In the future, this research may also lead to finding new medicines.

NATIONAL GEOGRAPHIC

BECOME AN EXPERT

Matter: The Uses of Matter

Matter is everywhere.

Everything you see, feel, taste, and smell is made up of matter. Many forms of matter are found naturally in the earth. Matter that is found in the earth is used to make many different things. Cars, video games, and even artificial body parts all began as matter from the earth.

Matter is found in the earth in all three states: solid, liquid, and gas. These materials are removed from the earth through mining and drilling. It takes well-trained people to decide whether or not to dig for matter. It is a difficult and expensive process.

These machines remove oil from the earth.

Energy like gasoline, heating gas, and electricity come from natural resources in the earth. Natural resources are things like trees, water, coal, oil, natural gas, and soil. We find natural resources in the environment and use them to create useful products.

Natural gas flows through these huge pipes.

Solids Many solids occur naturally in the earth. They are found by themselves or as mixtures of different materials. Iron, silver, and gold are all metals that are found in the earth. Sometimes the gold you find in Earth is not pure gold. It's part of a **mixture** called ore.

Gold is solid.

Falling liquid water creates energy.

Liquids Water is the most plentiful liquid on Earth. It covers 70 percent of Earth's surface. We use water for drinking, irrigation, and energy. Oil is another liquid pulled from earth.

Gases Natural gas comes from deep below Earth's surface. Natural gas is one of the most used forms of energy on Earth. Natural gas is made up mostly of methane. Methane is a compound that contains one **atom** of carbon and four atoms of hydrogen. The air in our atmosphere is also a resource that is being used to create energy. Wind power can generate electricity.

People harvest blowing air, a solution of gasses, to make energy.

mixture

A **mixture** is two or more kinds of matter put together.

atom

An **atom** is the smallest piece of matter that can still be identified as that matter.

Coal Fossil fuels started as the remains of plants and animals. Over millions of years, heat and pressure broke the remains down into much of what we use today for energy. There are three kinds of fossil fuels: coal, oil, and natural gas. Coal is found in the earth as a solid. It is a rock made up mostly of carbon, hydrogen, and oxygen. Of the three types of fossil fuels, coal is the most plentiful and least expensive. People around the world use coal to make electricity. Mining is the process of removing coal from the ground.

Using coal to create energy has some risks. In areas where coal is stripped off the surface of the land, the land is sometimes destroyed. Mining coal and burning coal also create air pollution. The coal industry and national and state governments are working to make the level of pollution go down and to take better care of the land around mines. These practices can make coal mining less risky and dangerous.

U.S.A COAL-PRODUCING STATES

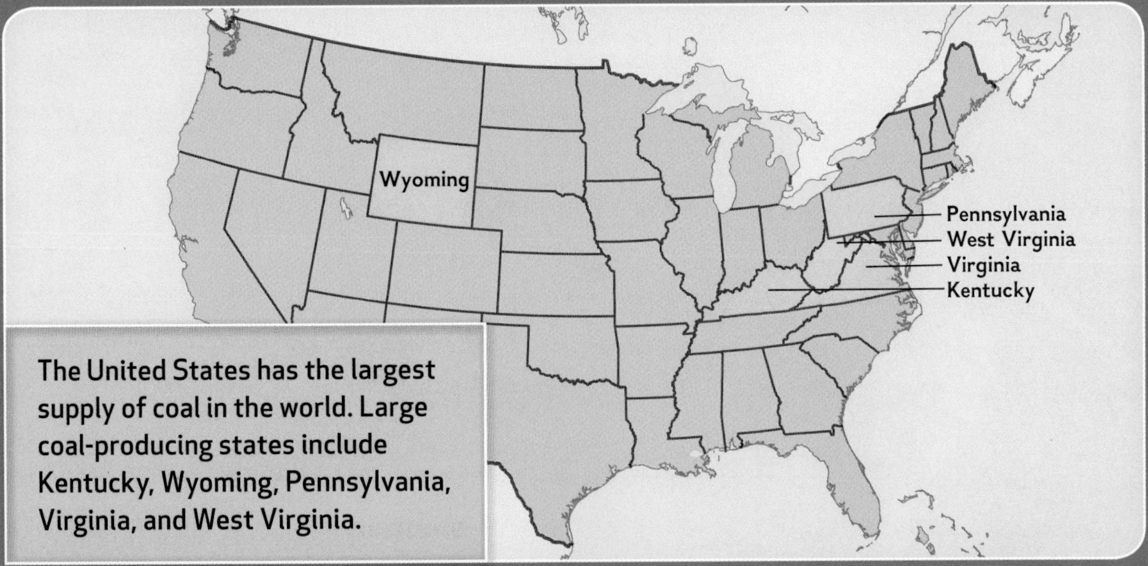

The United States has the largest supply of coal in the world. Large coal-producing states include Kentucky, Wyoming, Pennsylvania, Virginia, and West Virginia.

BECOME AN EXPERT

Diamonds Diamonds are a natural material. They are formed deep inside the earth. Diamonds have been discovered in more than 35 countries around the world. The **mass** of a diamond is measured in carats. The largest diamond ever mined was over 3,000 carats. It was discovered in 1905 in South Africa.

Gold Gold is a precious metal that is found in a solid state within the earth. Long ago, people could find gold by sifting through rocks in a river or hitting rocks with a pickaxe. Today, gold is found in much smaller amounts. To mine gold, miners blast large craters in the ground and remove huge amounts of ore. Ore is a mixture of gold and rocks, like quartz. The ore is crushed into a powder and mixed with water. Then, a chemical **solution** is added to dissolve the gold.

Almost half the world's gold can be found in South Africa.

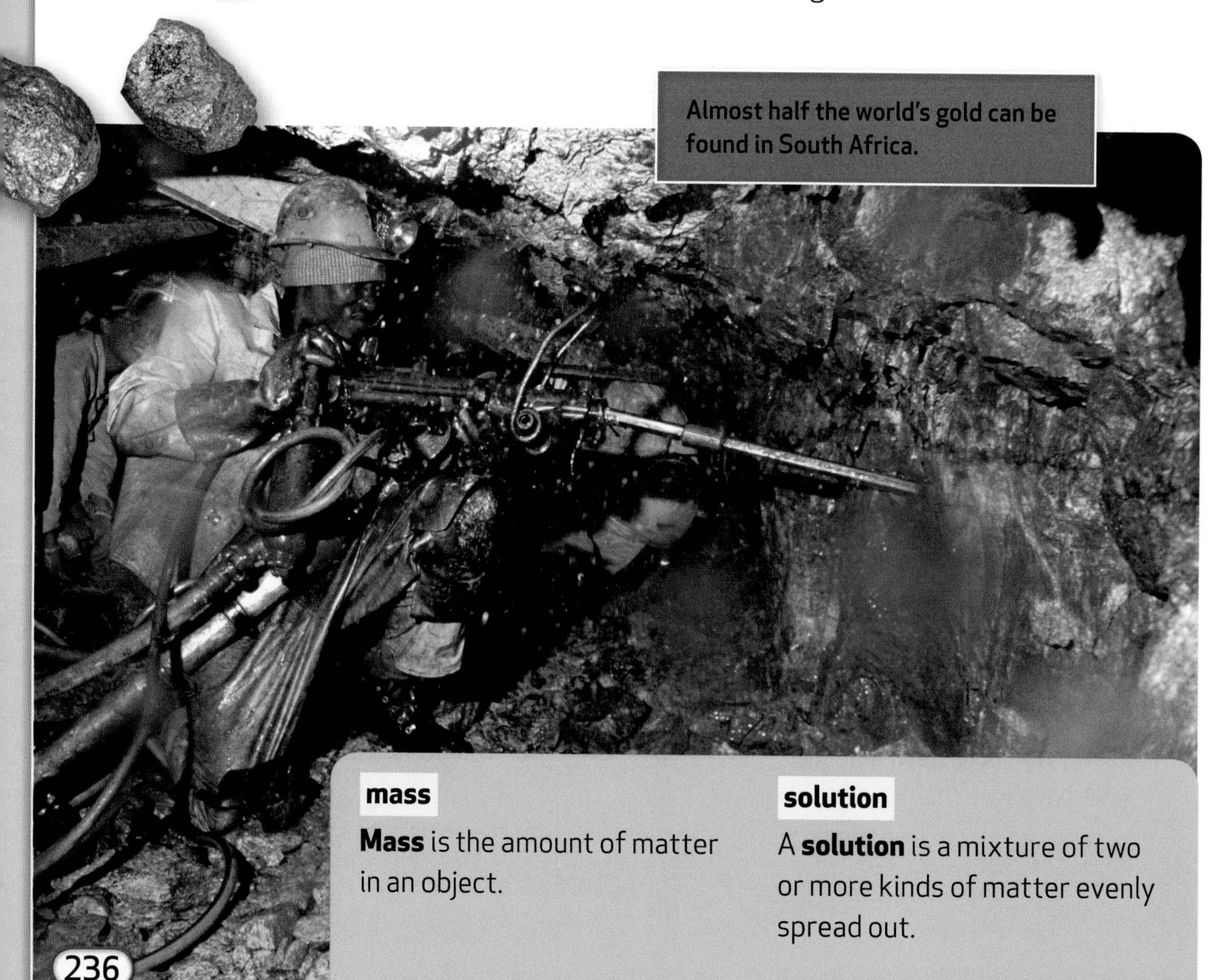

mass

Mass is the amount of matter in an object.

solution

A **solution** is a mixture of two or more kinds of matter evenly spread out.

Oil Oil, or petroleum, is a liquid fossil fuel found deep in the earth. Oil is a mixture made mostly of liquid carbon and hydrogen along with other materials. Geologists are scientists who study Earth. Geologists locate oil underground. After a site has been found and cleared, drilling begins. A well is drilled and prepared so no earth will fall into the hole. Then, a pump is placed in the well. An electric motor moves a lever up and down, creating suction. Like drinking with a straw, suction draws the oil up the pump.

Oil that is straight from the ground and has not been cleaned up or processed is called crude oil. Oil in the ground can have different colors and textures. Coal can be burned as it is when it comes out of the ground. Oil, however, must be refined before it can be burned. *Refined* mean that the oil has been cleaned up and the extra matter has been removed.

Oil wells create suction that draws oil from the earth.

BECOME AN EXPERT

Refining Oil Crude oil must be refined before we can use it. When crude oil, or petroleum, is refined, its many parts are separated with heating. The different materials in the oil boil at different temperatures. As each part boils, it changes to a gas. As each gas cools, it turns back into a liquid and is collected. We use each part of the oil for different things. The most common petroleum product is gasoline.

Some people estimate that there are over 6,000 products made from petroleum. The chart below shows just a few of the products made from petroleum.

A FEW PETROLEUM PRODUCTS

Lipstick

Sunglasses

Cleats

Ink

Crayons

Roller-skate wheels

Prudhoe Bay, Alaska The Prudhoe Bay oil field in Northern Alaska is the largest oil field in North America. Located on the coast of the Arctic Ocean, the ground is permanently frozen several feet deep in this bitterly cold place. The main Prudhoe Bay oil field has an area of about 1,554 square kilometers (600 square miles). There are smaller oil fields near the main one.

The Prudhoe Bay oil field is the largest oil field in North America.

ARCTIC OCEAN
RUSSIA
CANADA
Alaska
Bering Sea
Gulf of Alaska

Oil was discovered in the rock under Prudhoe Bay in 1968. Five years later the U.S. government approved plans for a 1,287.5-kilometer (800-mile) pipeline to carry oil from Prudhoe Bay to the port town of Valdez, Alaska. From Valdez, ships carry oil all over the world.

The 19 oil fields in Prudhoe Bay have produced more than 12.8 billion barrels of oil since 1977. Drilling and transporting oil affects the environment. Since 1995, there have been about 400 spills per year from the oil field and pipeline. The total **volume** of oil spilled is about 1.5 million gallons (5,678,118 liters). These spills cause damage to soil and death or injury to wild animals.

TECHTREK
myNGconnect.com
Prudhoe Bay, Alaska
Digital Library

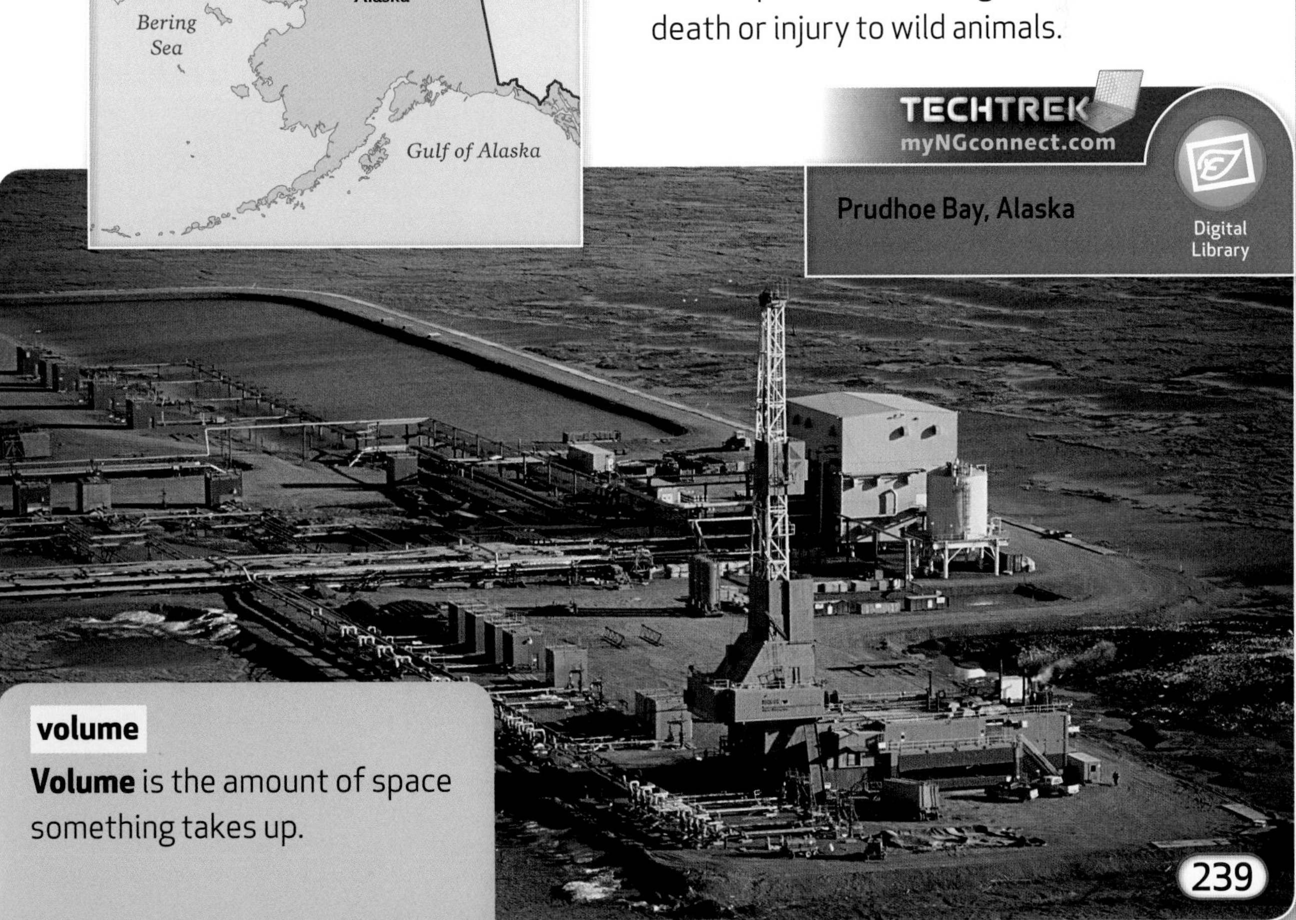

volume

Volume is the amount of space something takes up.

Natural Gas Natural gas is the third type of fossil fuel found in the earth. It occurs naturally in the gas state. Natural gas also has no color or smell. We often add a smell to natural gas so that we can smell it and avoid being near harmful fumes. Unlike some gases, natural gas burns easily. Natural gas is a very important source of energy throughout the world. We use it for heating homes, cooking, and creating electricity.

Natural gas is made of a mixture of carbon and hydrogen called methane. It also contains small amounts of other substances. Natural gas found in the earth must be refined much like oil before it can be used. Refining natural gas means cleaning out extra substances such as water, oil, gases, sand, and other compounds. After being refined, some of the parts of natural gas are sold separately. Propane (used in grills) and butane (used in lighters) are gases that are refined from natural gas.

The flame of a gas stove is fueled by natural gas.

Drilling for Natural Gas

Geologists locate natural gas in the earth. Once a site has been found, it the drilling company works to collect gas. There are two types of drilling used to get natural gas from the earth. One method is to drop a heavy metal bit into the ground over and over. This is called percussion drilling. This process creates a hole from which we can collect natural gas.The other method is called rotary drilling. A drill bit similar to one you may have at home rotates as it digs into the earth.

If the drilling company finds natural gas at a drilling site, workers begin the process of bringing the gas to the surface. First, workers strengthen the wall of the well. Then, they fit pipes and tubes to the top of the well to collect the gas. Sometimes the gas will rise up the tubing on its own. After all, gases expand to fit their containers. In other cases, pumps may need to pull the natural gas from the earth.

It takes special equipment to drill once natural gas is located.

This series of pipes at the top of a natural gas well is known as a Christmas tree. It controls the flow of matter out of the well.

Natural Gas Under the Sea

At the beginning of 2009, a natural gas deposit was found off the coast of Israel. In a country with few natural energy sources, this was a very big discovery. This was the largest natural gas site ever found in Israel. The site of this natural gas supply? Beneath the ocean floor! After drilling more than 4,572 meters (15,000 feet) into the sea floor, the gas deposit was confirmed. The energy company believes that there might be enough natural gas in this site to last 20 years. They plan to start bringing the gas to the surface in 2012.

The Gulf of Venezuela was also the site of a new natural gas discovery in 2009. Venezuela's state oil company believes that this could be one of the largest natural gas deposits in the world. It could contain seven to eight billion cubic feet of gas.

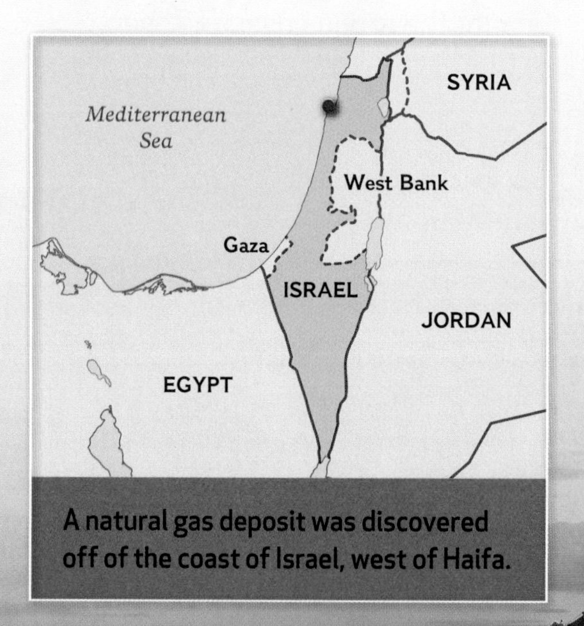

A natural gas deposit was discovered off of the coast of Israel, west of Haifa.

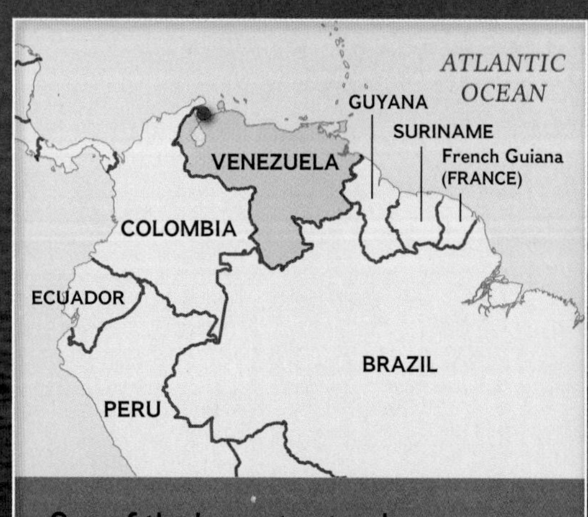

One of the largest natural gas deposits in the world was discovered beneath the Gulf of Venezuela.

Energy and the Future

We pump and dig different kinds of matter from the earth to make our energy. But most of this matter is nonrenewable. This means that it takes a long time for the matter to form. Once we use that matter, it's gone. At least, it's gone for millions of years! What other kinds of matter can we use for energy?

One type of matter we can use for energy is the air. Wind power can create electricity. Wind is simply the movement of air. Windmills capture the energy of wind and turn it into electricity. Wind energy is a renewable resource.

Water is a liquid form of power. Hydropower can create electricity. Water that flows swiftly creates a great deal of energy. That energy pushes against turbines. The moving turbines can create electricity.

Earth is rich with resources, matter that we use everyday. Taking care of Earth and its resources is one way to preserve that matter for use in the future.

CHAPTER 5

SHARE AND COMPARE

Turn and Talk How does the matter that we can find in the Earth affect our daily lives? Form a complete answer to this question along with a partner.

Read Select two pages in this section. Practice reading the pages. Then read them aloud to a partner. Talk about why the pages are interesting.

my SCIENCE notebook

Write Write a conclusion that tells the important ideas you learned about natural resources. State what you think is the Big Idea of this section. Share what you wrote with a classmate. Compare your conclusions. Did your classmate come to the conclusion that natural resources exist in all three states of matter?

my SCIENCE notebook

Draw Draw a picture that represents the origin of a fuel that comes from Earth. Add a caption to your drawing. Then compare your drawings to those of other classmates. Create a display with all your drawings to show the various ways that fuels come from Earth.

In Chapter 6, you will learn:

FLORIDA NEXT GENERATION SUNSHINE STATE STANDARDS

SC.5.P.13.1 Identify familiar forces that cause objects to move, such as pushes or pulls, including gravity acting on falling objects. **FORCE AND MOTION**

SC.5.P.13.2 Investigate and describe that the greater the force applied to it, the greater the change in motion of a given object. **NEWTON'S LAWS OF MOTION**

SC.5.P.13.3 Investigate and describe that the more mass an object has, the less effect a given force will have on that object's motion. **NEWTON'S LAWS OF MOTION**

SC.5.P.13.4 Investigate and explain that when a force is applied to an object but it does not move, it is because another opposing force is being applied by something in the environment so that the forces are balanced. **NEWTON'S LAWS OF MOTION**

SC.5.P.13.4 Science in a Snap! Investigate and explain that when a force is applied to an object but it does not move, it is because another opposing force is being applied by something in the environment so that the forces are balanced.

CHAPTER

6

HOW DO YOU DESCRIBE FORCE AND THE

Things are in motion all around you. Every motion that you see is the result of forces. At the circus, you may see people flying through the air during a trapeze act. The acrobats swing, hold, and twist. They use force in each move to perform the routine.

Student eEdition | Vocabulary Games | Digital Library | Enrichment Activities

The acrobat uses force to move.

LAWS OF MOTION?

SCIENCE VOCABULARY

force (FORS)

A **force** is a push or a pull. (p. 250)

Forces acting on a object can cause the object to move.

motion (MŌ-shun)

Motion is a change in position. (p. 250)

The ball and some of the cans are in motion.

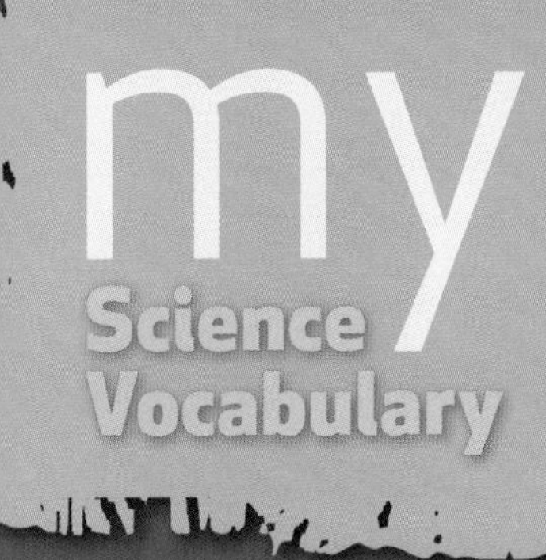

acceleration
(ak-sel-er-Ā-shun)

force
(FORS)

gravity
(GRA-vi-tē)

motion
(MŌ-shun)

gravity (GRA-vi-tē)

Earth's **gravity** is a force that pulls things to the center of Earth. (p. 252)

Gravity pulls the girl into the water.

acceleration (ak-sel-er-Ā-shun)

Acceleration is a change in an object's velocity. (p. 261)

An object's acceleration depends on the mass of the object and the size of the force acting on it.

Force and Motion

Look at the children below. They are taking their dog to the park. They are pulling him in a wagon. The children are using a **force** to make the wagon move. Force is a push or a pull that causes an object to move.

How can you tell if something is moving because of a force? Look at the object. Is it in a new place?

Think about the wagon that the children pulled. Before they pulled the wagon, it was still. After they used force, the wagon moved. **Motion** is a change in position. When the wagon moved to a new position, it was in motion.

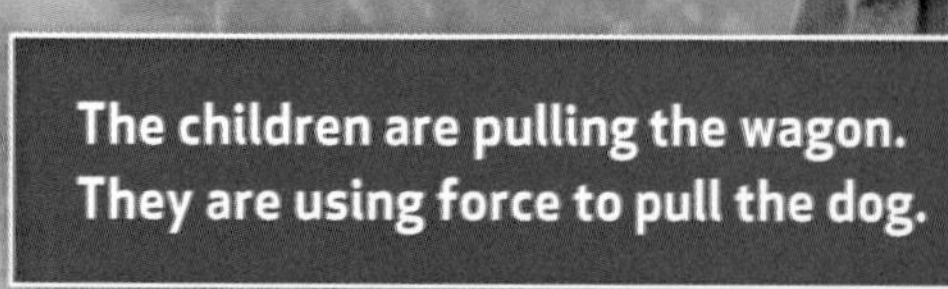

The children are pulling the wagon. They are using force to pull the dog.

Direction Direction is one way to describe motion. You can use compass directions, such as north and south. Or you can use many other directions. The bird flew up. The car turned left. Objects can also move toward and away. The rabbit ran toward the gate. The dog ran away from the house. Direction helps describe where an object is going.

Direction also helps to describe forces as well. In the picture, the children are pulling the wagon. The wagon is moving toward the children. So, the direction of the force being exerted on the wagon is toward the children.

Gravity Suppose you drop a ball. The force that causes the ball to fall to the ground is **gravity**. The gravity between Earth and the ball causes the ball to move to the center of Earth. This happens even though Earth and the ball are not touching each other. Gravity pulls objects downward unless something holds them up.

Although the Earth is not touching this swimmer, she is pulled toward Earth by Earth's gravity.

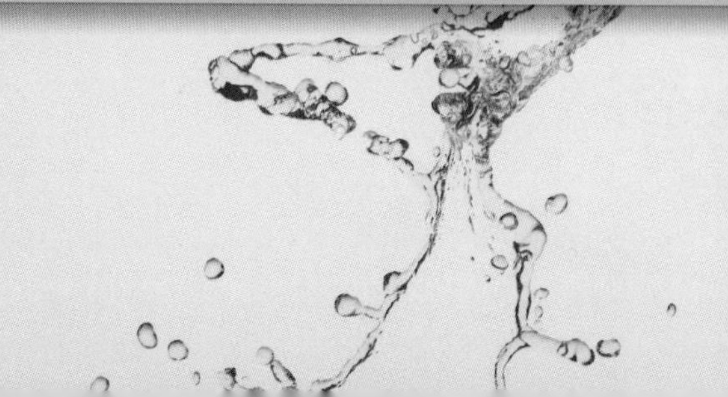

Look at the four photos of the girl. She is jumping into a lake. Have you ever jumped into the water? You may stand on a platform near the edge of the water. Then you jump! What happens? You land in the water and make a big splash.

Why did this happen? Earth's gravity is pulling you to the center of Earth. When you jump off the platform, Earth's gravity pulls you downward into the water. Earth's gravity is a force that pulls on everything on Earth.

Before You Move On

1. What are the two types of forces?
2. What does gravity do to objects on Earth?
3. **Predict** Imagine kicking a ball in the air. What does the ball do? Why?

NATIONAL GEOGRAPHIC

SIR ISAAC NEWTON
GRAVITY AND THE LAWS OF MOTION

Isaac Newton was one of the greatest scientists of all time. He was born in 1642 in England. When he was 16, his mother took him out of school to begin working on the farm. However, Newton did not like being a farmer. And so, when he was 18, he went to college.

Newton studied math, physics, and astronomy. While he was there, he had to work as a servant to pay for college. After he graduated, he stayed to work as a teacher. But later that year, the Great Plague broke out, and the school was closed. Newton had to go back home to the family farm.

Isaac Newton made many important contributions to science. Not only did he describe the laws of motion and of gravity, he also built a telescope using mirrors, developed a new kind of math, and studied light and astronomy.

After he returned home, Newton continued learning on his own. Some of Newton's greatest discoveries happened during this time. One such discovery occurred while observing the effects of gravity.

Newton realized that gravity is a force. It caused objects to fall to Earth. He realized that the same force must also keep the moon in orbit. Newton discovered that gravity depends on the mass of objects and the distance between them. He worked out an equation to find the strength of gravity. Scientists still use his equation today!

Newton did not publish his ideas about gravity until 1687. These ideas were published in three books. Together, they are called *Principia*. The *Principia* includes some of the most important ideas ever written about science.

MATHEMATICAL

PRINCIPLES

OF

NATURAL PHILOSOPHY.

By Sir ISAAC NEWTON, Knight.

TRANSLATED INTO ENGLISH, AND ILLUSTRATED WITH A COMMENTARY,

By ROBERT THORP, M.A.

VOLUME THE FIRST.

LONDON:

PRINTED FOR W. STRAHAN; AND T. CADELL, IN THE STRAND.

MDCCLXXVII.

The first book of *Principia* begins with a description of three laws of motion. These three laws are now called Newton's laws of motion. The laws describe how objects move when acted on by forces. Each law is listed in the table below.

The third book of *Principia* describes how mass and distance affect the gravity between two objects. The third book also explains how gravity affects the motion of the planets and other bodies in the solar system.

By writing the *Principia*, and making many other important discoveries, Newton made many important contributions to science.

Newton discovered that sunlight could be separated into colors.

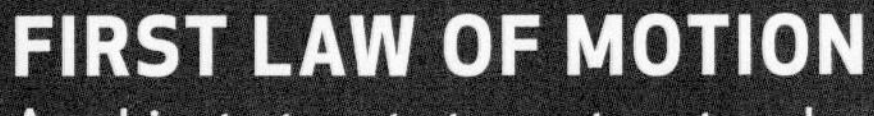

FIRST LAW OF MOTION

An object at rest stays at rest and an object in motion stays in motion at the same speed and in the same direction unless acted on by an unequal force.

SECOND LAW OF MOTION

The acceleration of an object depends on the mass of the object and the size of the force acting on it.

THIRD LAW OF MOTION

For every action, there is an equal and opposite reaction.

Newton's Laws of Motion

Look at the girls on the slide. Many different forces are acting on the girls. The water is pushing them down the slide. Gravity is pulling them down the slide. The forces that are moving the girls down the slide are greater than the forces that are holding them up. Because of these unequal forces, the girls slide down into the pool.

Sometimes different forces cancel each other out. Suppose you and your friend are both pulling on a book. You notice that even though you and your friend are pulling really hard, the book doesn't really seem to change position. This is because the forces that you and your friend are applying to the book are the same and cancel each other out. Unlike the forces that are pulling the girls down the slide, the forces that you and your friend are exerting on the book are equal, and so the book does not move.

The force of gravity pulls the girls down the slide.

These cans will stay at rest unless they are acted upon by a greater force, such as the moving ball.

Newton's First Law Any object that is not moving is said to be at rest. Many of the cans in the picture are at rest. According to Newton's first law, an object will stay at rest unless acted on by a greater force. In other words, something that is not moving will not start moving until it is pushed or pulled. When the cans are hit by the ball or other falling cans, they will move.

The law also states that an object will continue in the direction and at the speed that it is moving unless acted on by a greater force. After the ball hits the cans, it may hit a wall and then stop. The wall exerts a greater force on the ball than the ball exerted on the wall, and so the ball stops moving.

Newton's Second Law Look at the photos below. Which person is using more force? How can you tell?

You know that the Newton's first law of motion tells you that a push or a pull can change the motion of an object. His second law of motion tells you that a harder push or pull will cause a greater change in the object's motion. It also tells you that an object with more mass is harder to move than an object with less mass.

In the photos below, the big splash shows a greater change in the water's motion. It shows a person using a lot of force to make a large amount of water move quickly into the air. The small splash in the second picture, shows less force being used.

Another important part of Newton's second law of motion is **acceleration**. Acceleration is a change in an object's velocity, which is an object's speed and direction.

An object's acceleration depends on the mass of an object and the size of the force acting on it. Suppose you were throwing a baseball. The more force you exert, or the speed at which you move your arm, affects the speed at which the baseball moves.

However, if you tried to throw a bowling ball, you would to need a use a lot more force than when throwing the baseball. As the object's mass increases, so does the amount of force needed to accelerate the object. Since the bowling ball has more mass than the baseball, it is harder to throw at the same speed.

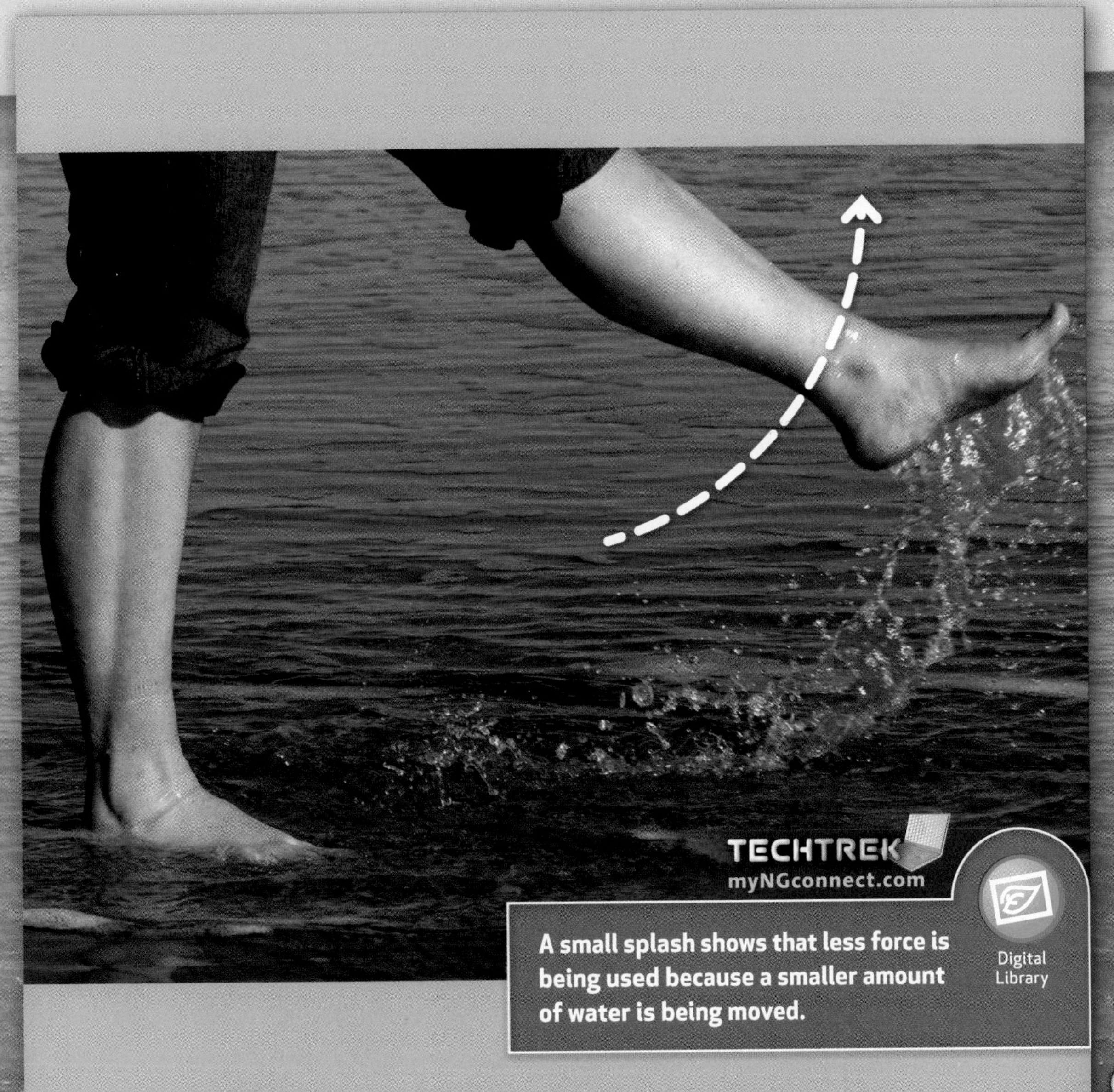

A small splash shows that less force is being used because a smaller amount of water is being moved.

Newton's Third Law Look at the picture below. How does the woman get the kayak to move? Newton's third law of motion helps explain how she does it.

When she paddles the kayak, she pushes the paddle against the water. The water pushes back against the paddle. The two forces exert in opposite directions. The result of the forces is that the kayak moves forward.

Newton's third law is often summarized as, "For every action, there is an equal and opposite reaction." When the woman forces the paddle through the water, which is her action, the water pushes back against the paddle with an equal and opposite reaction, and off she goes!

Sometimes the equal and opposite forces cause no change in motion. For example, the people that are sitting at the beach are exerting force on their chairs. The chairs are exerting an equal and opposite force on the people. The result of these forces is that the people are seated in their chairs and do not fall down. The forces being applied are the same.

When you are sitting, you push down on the chair. The chair pushes up on you.

Science in a Snap! Balloon Rocket

Blow up a balloon. Tape a straw on top of the balloon while you hold the end of the balloon.

Thread a string through the straw. One partner holds each end of the string. Move the balloon to one end of the string. Let go of the balloon.

What happened to the balloon? How does the third law of motion explain what you observed?

Before You Move On

1. Why is it easy to push a toy truck, but hard to push a real truck?
2. How does Newton's third law of motion explain how to move a skateboard that you are standing on?
3. **Apply** How does a bed exert force on you while you are lying in it?

Conclusion

Motion can be described by an object's direction and speed. Motion is caused by forces. Changes in motion are caused by unequal forces. The object's acceleration determines the amount of change in an object's position. Acceleration depends on the object's mass and the force acting on it.

Big Idea Newton's laws of motion describe how the forces on an object cause changes in the object's motion.

Vocabulary Review

Match the following terms with the correct definition.

A. force
B. gravity
C. motion
D. acceleration

1. a push or a pull
2. a change in an object's velocity
3. a force that pulls things to the center of Earth
4. a change in position

Big Idea Review

1. **Explain** How is force related to each of Newton's laws of motion?

2. **Infer** A child and an adult are ice skating. The child pushes away from the adult. The child moves backward. What will happen to the adult?

3. **Predict** A feather falls off a bird in flight. The force of gravity acts on the feather. How does this force affect the feather's motion?

4. **Infer** Use Newton's first law of motion to explain why it is safer to wear a seatbelt in a moving car.

5. **Develop a Logical Argument** Use Newton's second law of motion to explain why kicking a bag of soccer balls across a field takes more force than kicking one soccer ball across a field.

6. **Analyze** Two forces are acting on an object at rest. How can you tell if the forces are equal or unequal?

Write About the Laws of Motion

Apply Concepts Think about a sport or an activity that you do at home. Describe how each law of motion is observed in the sport or activity.

What Does a Kinesiologist Do?

William Sands is a kinesiologist. He uses his knowledge of forces, motion, and the human body to help top athletes and others perform at a higher level.

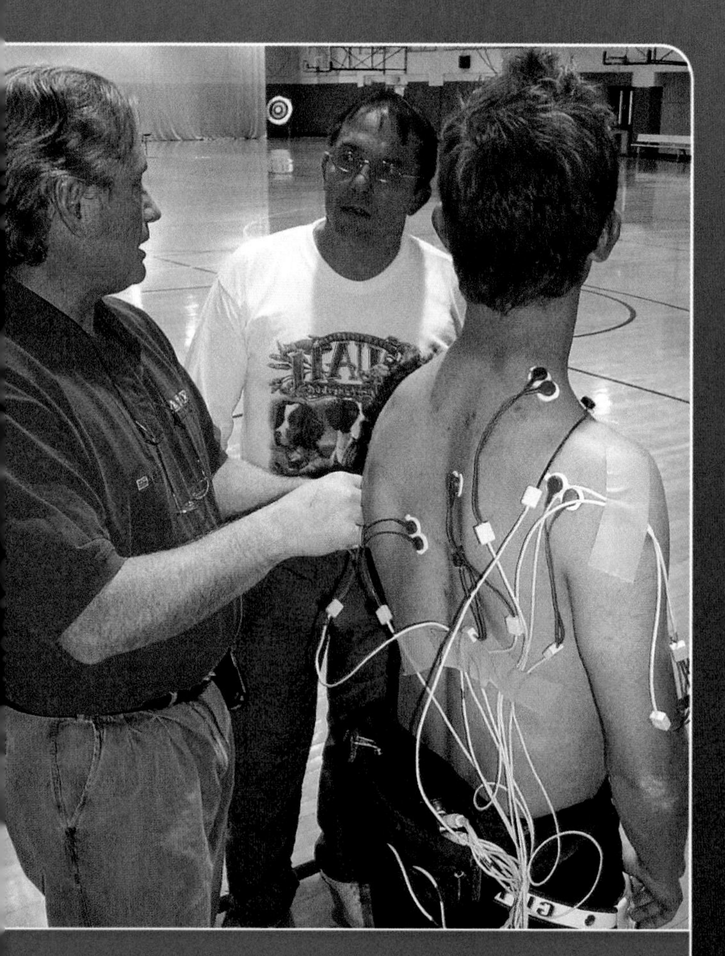

William Sands connects the archer to a computer to see if the archer is moving the muscles in his body correctly.

NG Science: What do you study?

William Sands: Most of my research time is spent on studying elite athletes, specifically gymnasts. I study just about all aspects of high performance in order to make better athletes. I analyze where the athlete makes mistakes in their routine. This helps me make the athlete's performance better in the future.

NG Science: What type of research have you done?

William Sands: I research the human body and the way it moves. For example, in the picture to the left, I am putting electrodes on a national champion archer. These electrodes are used to detect tiny electrical signals that are sent by his brain to the muscles in his arms which produces force. I study these tiny signals to determine just how elite athletes use their muscles to accomplish their amazing skills.

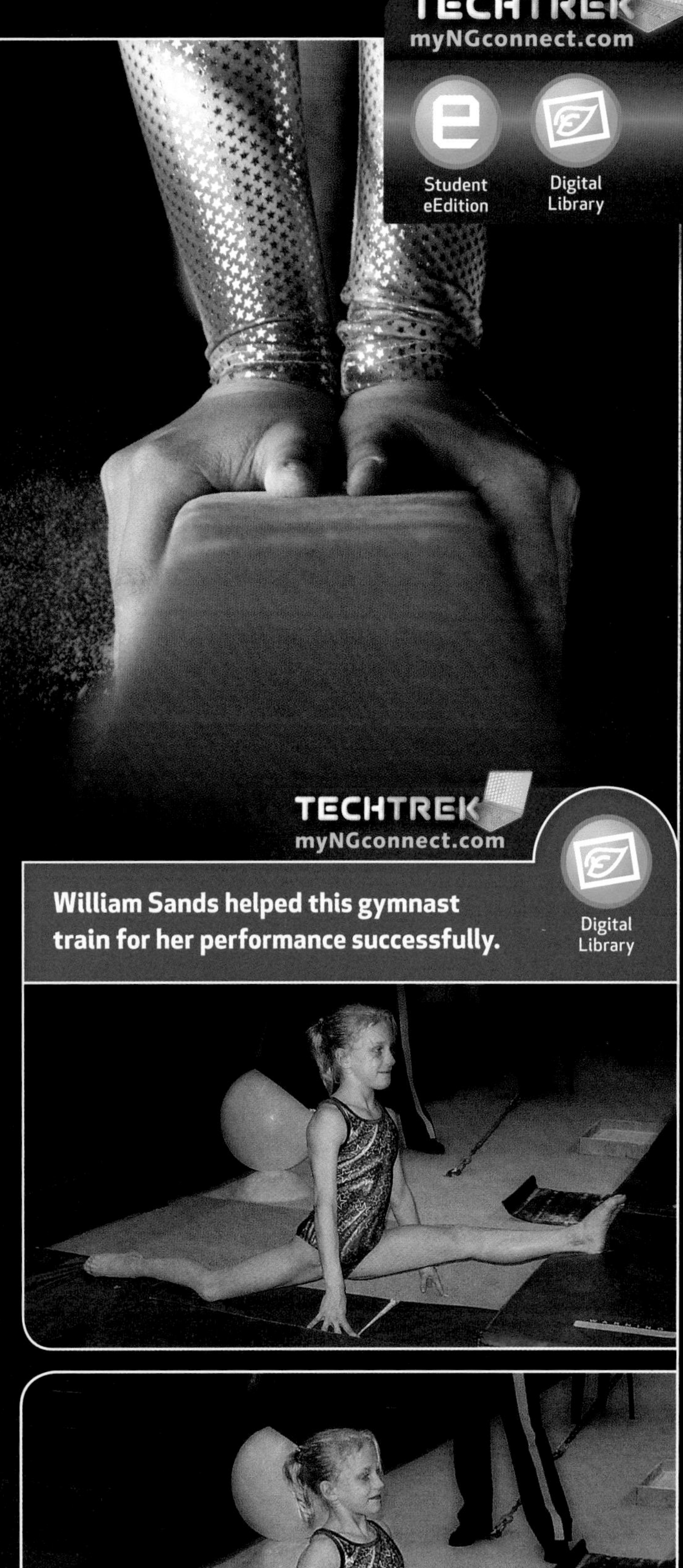

TECHTREK myNGconnect.com

Student eEdition

Digital Library

NG Science: What is your favorite thing about your area of research?

William Sands: The thrilling part of my research is when I see the athletes that I work with achieve their goals. I use science to help them, but the science is small compared to their dedication, determination, and commitment to their sport.

NG Science: What have you enjoyed most during the course of your career?

William Sands: I was lucky enough to coach several Olympians in women's gymnastics. I have now turned to science to help even more people with movement problems.

NG Science: What type of training do you need to become a kinesiologist?

William Sands: To become a kinesiologist, you will need to study many kinds of science, as well as mathematics. You will also need to choose the type of body motion you find most interesting.

NG Science: Why is kinesiology important?

William Sands: If it moves, kinesiologists study it. Kinesiologists study motion so that everyone can move well; whether you want to dunk a basketball or simply rise from a chair.

TECHTREK myNGconnect.com

Digital Library

William Sands helped this gymnast train for her performance successfully.

NATIONAL GEOGRAPHIC

BECOME AN EXPERT

Gymnastics: Forces in Action

Athletes in every sport learn how to exert the right **forces** to do well. In sports such as gymnastics, athletes use forces to control their own bodies.

Gymnasts use forces to perform different skills. Their hands and feet push off from the ground when doing tumbling. Their arms pull on the uneven bars and the parallel bars. As they exert these forces, each of the gymnasts are in **motion**. The movement of their bodies follows Newton's laws of motion.

Like all sports, gymnastics involves exerting the right forces at the right times.

force

A **force** is a push or a pull.

motion

Motion is a change in position.

Some gymnastics skills require a gymnast to stay in one position. Men on the still rings perform strength moves. To keep his body in the correct position, a gymnast has to make sure that the forces on his body are equal.

Gravity pulls on the gymnast while he is on the still rings. What happens if the gravity and other forces on the gymnast become too strong? If the gymnast pushes upward too hard or not hard enough, his body will not move in the correct way.

To hold this position, this gymnast needs to keep the forces on his body equal.

gravity

Earth's **gravity** is a force that pulls things to the center of Earth.

The First Law of Motion in Gymnastics

Before a gymnast starts moving, he or she is at rest. To start moving, the gymnast has to exert an unequal force. According to Newton's first law of motion, an unequal force will cause an object at rest to move. Often, that unequal force is the gymnast's feet pushing off the ground. But sometimes that force is a pull. Many times, men have to pull themselves up on the different apparatus to start their routines.

This gymnast had to exert unequal forces to move his body into the handstand position. But once he got into the position, he had to keep the forces equal so that he would not fall.

An object in motion will stay in motion unless acted on by an greater force. You can see this part of Newton's first law whenever a gymnast dismounts from the uneven bars. Right before a gymnast dismounts, she pushes off from one of the bars. Then, she starts flying through the air. While she is in the air, gravity is pulling her down. She stops moving when she lands on the ground. The ground's force is equal to the force exerted by the gymnast.

When a gymnast is ready to dismount, she must push off from one of the bars.

After the gymnast pushes away from the bars, Earth's gravity is pulling her to the ground.

The Second Law of Motion in Gymnastics

When a gymnast wants to perform a small hop on the beam, she exerts only a small force. But when the gymnast wants to dismount off the beam, she exerts a large force. She jumps high in the air. She is high enough to do twists or somersaults in the air before she lands.

How does the second law of motion help to explain this? To perform different moves, gymnasts have to exert different amounts of force.

TECHTREK myNGconnect.com

Digital Library

A gymnast exerts forces of different sizes to perform different skills on the balance beam.

Acceleration depends on force and mass. Remember, gymnasts create force by using their muscles. You have to be strong to be a gymnast. It doesn't matter how big or small you are , a lot of force must be exerted to perform these very difficult movements.

A gymnast exerts force to balance on the beam.

A gymnast must have enough acceleration to complete a dismount.

Acceleration

Acceleration is a change in an object's velocity.

The Third Law of Motion in Gymnastics

Let's look at Newton's third law of motion in action. The third law states that for every action there is an equal and opposite reaction. Look at the picture below. The gymnasts exert forces on the equipment. At the same time, the equipment pushes back on the gymnasts. These forces push them up into the air. Without these forces, gymnasts couldn't perform.

The pommelhorse pushes on the gymnast with an equal and opposite force that keeps the gymnast's body in the air.

These forces are equal in size. They are also opposite in direction. Look at the picture below. A gymnast pushes on the vault. The vault pushes back with a force of equal size on the gymnast. This force sends the gymnast flying up off the vault.

Gymnasts train their bodies to move in precise ways.

NATIONAL GEOGRAPHIC

CHAPTER 6 SHARE AND COMPARE

Turn and Talk How do equal and unequal forces affect the way gymnasts perform their routines? Form a complete answer to this question together with a partner.

Read Select two pages in this section. Practice reading the pages. Then read them aloud to a partner. Talk about why the pages are interesting.

Write Write a conclusion that summarizes what you have learned about the different ways gymnasts use force. In your conclusion, restate what you think is the Big Idea of this section. Share what you wrote with a classmate. Compare what each of you wrote.

Draw Forms groups of three. Have each person draw a picture that shows one of Newton's three laws of motion. Write labels that explain your drawing. Compare your drawing with your partners' drawings.

In Chapter 7, you will learn:

FLORIDA NEXT GENERATION SUNSHINE STATE STANDARDS

SC.5.P.10.1 Identify some basic forms of energy, including light, heat, sound, electrical, chemical, and mechanical. **ENERGY, MECHANICAL ENERGY, SOUND, LIGHT, HEAT, CHEMICAL ENERGY**

SC.5.P.10.2 Investigate and explain that energy has the ability to cause motion or create change. **ENERGY, MECHANICAL ENERGY, CHEMICAL ENERGY**

SC.5.P.10.1 Science in a Snap! Identify some basic forms of energy, including light, heat, sound, electrical, chemical, and mechanical.

CHAPTER 7

HOW DO YOU DESCRIBE DIFFERENT OF EN

Energy is all around you. You have energy when you move or even talk. Energy comes in many different forms. Motion and sound are two forms of energy. There are other forms, too. Light energy helps you to see. Heat energy keeps you warm. The food that you eat contains chemical energy.

Race cars use the energy stored in fuel.

TECHTREK
myNGconnect.com

Student eEdition

Vocabulary Games

Digital Library

Enrichment Activities

SCIENCE VOCABULARY

potential energy
(pō-TEN-shul EN-er-jē)

Potential energy is stored energy. (p. 284)

The blue car has more potential energy because of its position at the top of the ramp.

kinetic energy
(ki-NET-ik EN-er-jē)

Kinetic energy is the energy of motion. (p. 284)

The cars are in motion. Their potential energy turned into kinetic energy.

my Science Vocabulary

chemical energy
(KEM-i-kul EN-er-jē)

kinetic energy
(ki-NET-ik EN-er-jē)

mechanical energy
(mi-CAN-i-kul EN-er-jē)

potential energy
(pō-TEN-shul EN-er-jē)

mechanical energy
(mi-CAN-i-kul EN-er-jē)

The **mechanical energy** of an object is its potential energy plus its kinetic energy. (p. 284)

This toy's mechanical energy comes from its potential energy plus its kinetic energy.

chemical energy
(KEM-i-kul EN-er-jē)

Chemical energy is energy that is stored in substances. (p. 298)

Your body runs on the chemical energy stored in food.

Energy

Energy and work might make you think of a long Saturday afternoon raking leaves. You start out with energy. You do the work. And then you might feel as if you've run out of energy.

Because you had energy, you did work. All forms of energy can do work or cause change. Work is done when a force moves an object. You used your energy to cause change. What changed? You moved the leaves around. And you probably flattened some blades of grass while you raked. In other words, you caused a change in position or you moved things. Motion is the easiest kind of work to see.

You can see that this girl is doing work because you can see the change she is causing. She is moving the scattered leaves into a pile.

Every time you move, you use energy. Sometimes, you transfer some energy to something else. That's what this kicker is doing to this ball—he's transferring energy to the ball. That energy puts the ball in motion. The energy changes the ball's position. The ball moves. When it lands, it's going to do work. How? It's going to bend some blades of grass. Anything can have energy and do work.

When you kick a football, you give energy to it to make it move. You apply work to it.

Before You Move On

1. What is energy?
2. How are work and force related?
3. **Apply** You kick a soccer ball to a friend. Explain how you have transferred your energy.

Mechanical Energy

The cars on this ramp have stored energy which is also called **potential energy**. The cars have the ability to do work, but they aren't doing it at the moment. The blue car is higher off the ground than the red car. It has more potential energy because it has a farther way to travel.

When you let go of the cars, their potential energy turns into **kinetic energy**, or the energy of motion. All objects that are in motion have kinetic energy.

The **mechanical energy** of the cars is their potential energy plus their kinetic energy.

The blue car is higher than the red car. It has more potential energy.

The cars are in motion. Their potential energy turned into kinetic energy.

This toy also has potential energy. There is a spring inside. When you turn the knob, you tighten the spring. The tighter it gets, the more potential energy the toy has.

What happens when you release the knob? The spring starts to loosen. The potential energy turns into kinetic energy and the toy moves. Now it has energy of motion. The toy has some potential energy until the spring has fully loosened. The potential energy plus the kinetic energy is the mechanical energy of the toy.

Before You Move On

1. What is the mechanical energy of an object?
2. When does potential energy turn into kinetic energy?
3. **Infer** Look at the photos of the toy cars going down the ramp. When do you think the cars have the most kinetic energy?

Sound

Have you ever heard someone play the piano? The photo on this page shows the inside of a piano. A hammer strikes a set of wires, or strings. The wires vibrate. This vibration travels. Listeners then hear the sound of the piano.

Sound is another form of energy. It is energy you can hear. Sound travels in waves. Objects make sound waves when they vibrate. The waves travel away from the object. They spread out in all directions. They move through the air, liquids, and solids.

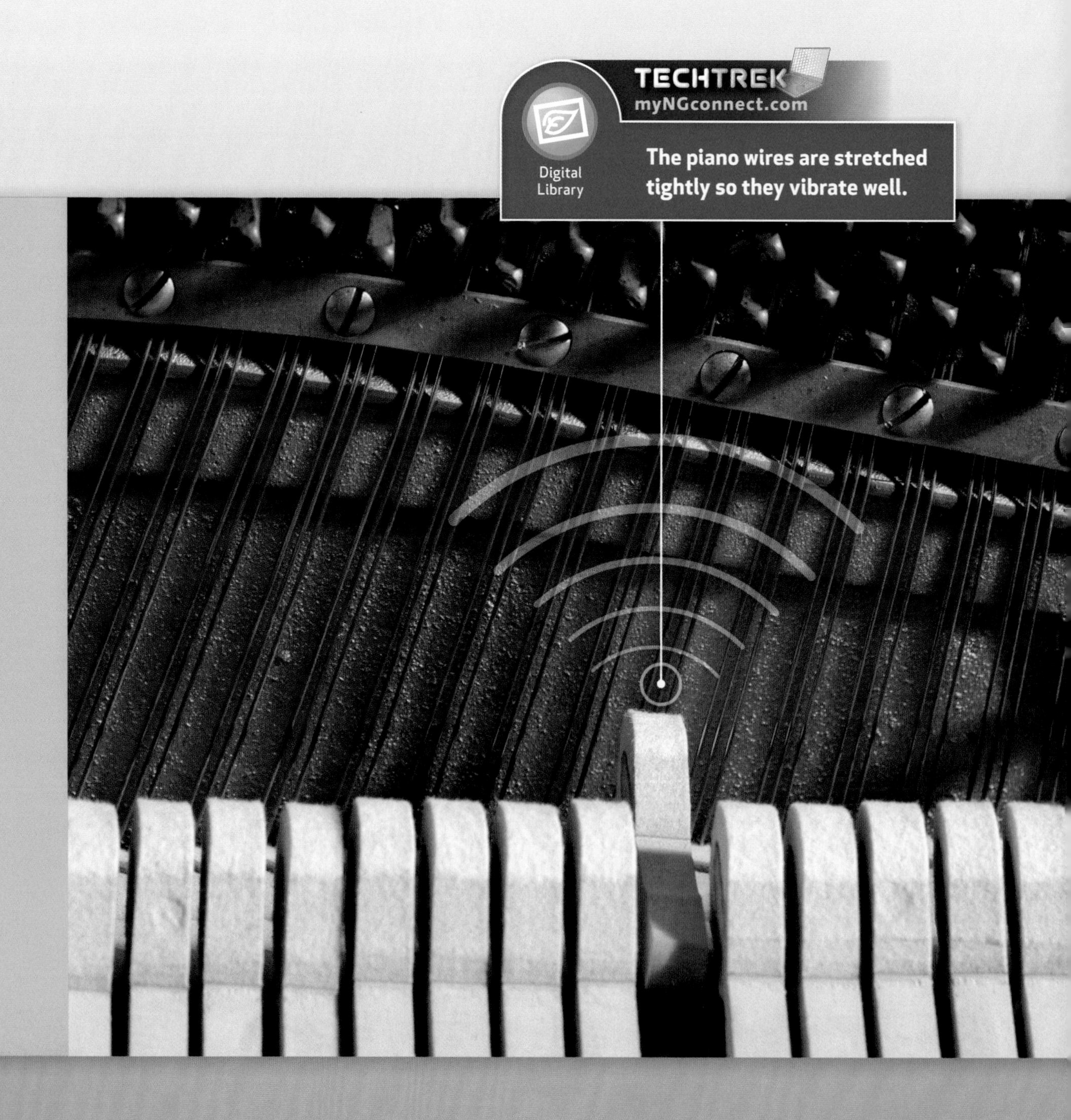

The piano wires are stretched tightly so they vibrate well.

A piano makes many different sounds. These sounds can have different pitches. Pitch describes how high or low a sound is. A piano makes high-pitched sounds and low-pitched sounds. It can make these different sounds because each wire is a little different.

Short, thin wires produce the piano's high-pitched sounds. Because the wires are short and thin, they vibrate quickly, so we hear a sound with a high pitch.

Thick, long strings produce the piano's low-pitched sounds. Because the wires are thick and long, they vibrate slowly to produce a sound with a low pitch.

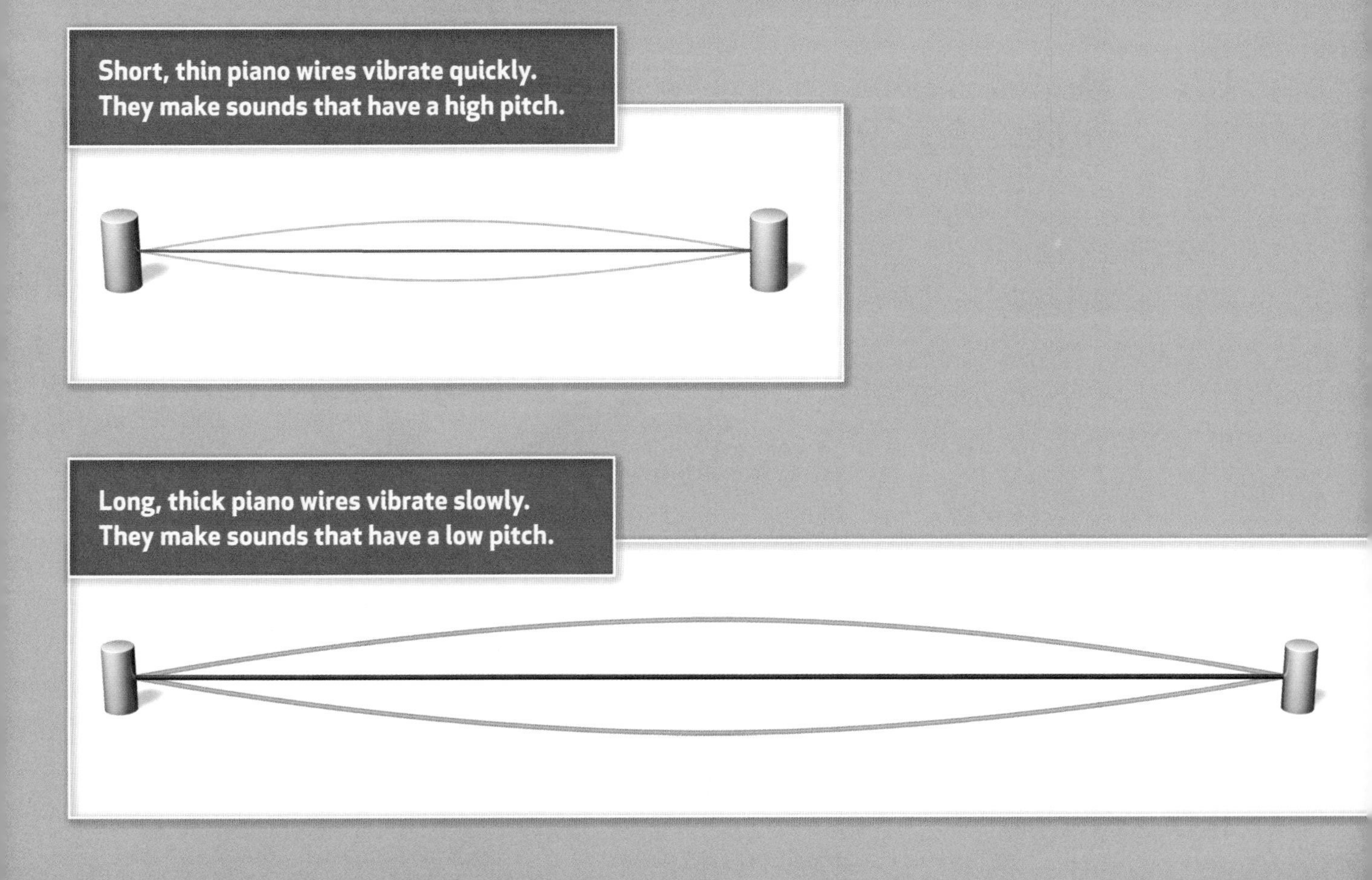

Reflection Some animals such as bats and dolphins use sound waves to find their way around and locate food. Look at the photo below. This bat is using sound waves to tell exactly where the moth is. The bat is only able to do this because sound waves reflect, or bounce off of, surfaces.

Echoes are reflected sound waves. You hear an echo when the reflected waves bounce back to your ears. In this picture, the sound waves reflect off the moth and back to the bat. The bat is using echoes to find his next meal.

TECHTREK
myNGconnect.com
Digital Library

The bat uses reflected sound waves to find food.

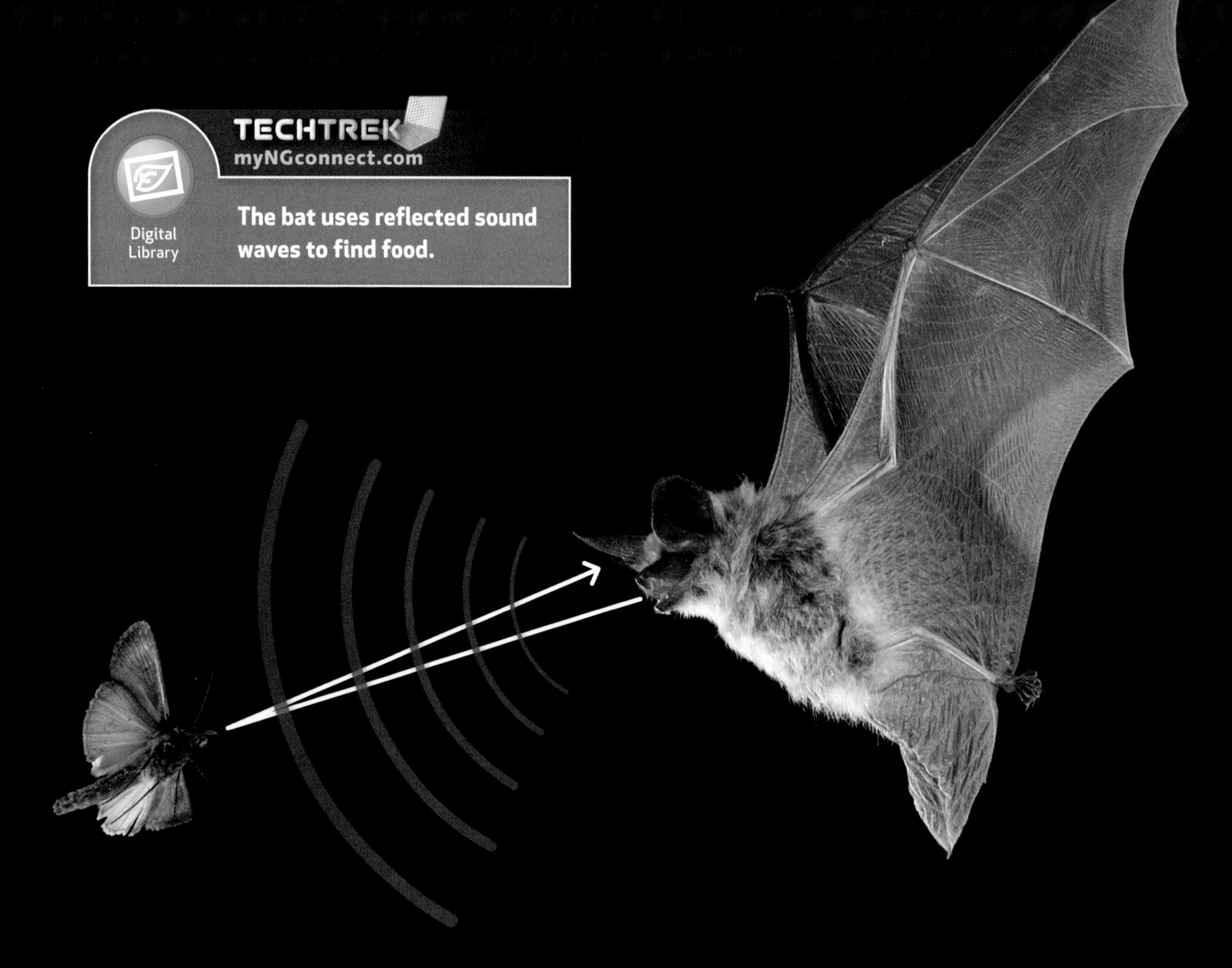

Absorption Think of a library. It's carpeted. It's full of books, which are made of paper and cardboard. Those materials absorb sound energy well. Sound waves do not bounce off of them. Instead, sound waves go into the materials. The materials absorb some of the energy, so the vibrations become weaker. Finally, the sound waves fade. Libraries seem quiet because the materials in the library are absorbing the sound waves instead of reflecting them.

Sounds waves are absorbed by materials in a library.

Before You Move On

1. What causes a sound wave?
2. A thick piano wire and a thin piano wire are vibrating. Compare the sounds they make, and tell why they are different.
3. **Generalize** Tell why a gymnasium would not be a good place to have a quiet library.

Light

Have you every noticed how one light bulb can light up an entire room? How does that happen? Light energy travels in waves. You can't see the waves, but you can see the light. Light travels away from its source in every direction. Light energy travels from the light bulb out to all parts of the room.

Light travels in a straight line. Think about being outside on a sunny day. The sun's light shines from above. But it can't shine around corners. So, there is a shadow beside a building. The light can't shine around you, either. So, you have a shadow too.

Light travels in a straight line. If something blocks the path of the light waves, such as your body or your hand, there is a shadow.

Reflection Look at the building in this photo. Based on your observations, what can you infer about the building's windows? The building's windows are mirrors. They reflect light. This means the light waves bounce off of them.

When light waves bounce back from a mirror, they make a reflection that you can see.

Because the building has so many mirrors on the outside, you can see a reflection of all the buildings across the street and of the sky.

Light waves from a nearby building reflect off of the mirrored building. They make a reflection.

Refraction What do you notice about the stem of this flower? It looks bent at the surface of the water. It looks bent because the light waves travel at one speed through the water, and at another speed through the air.

Light travels through clear materials. It bends, or refracts, when it passes from one material into another. It bends because the speed of light changes.

Light waves bend when they pass from the water to the air. The stem seems to have two parts because of the way the light waves bend.

Look at the prism on this page. When light passes through a prism, different wavelengths of light are bent by different amounts. When light enters the prism and then exits, the different colors of light bend by different amounts. This bending causes the colors of light to separate. That is why you see a rainbow when the light exits the prism.

Science in a Snap! Light Illusion

Pour water into a clear cup. Place a straw in the water.

Look at the straw through the side of the cup.

Describe what you see.

Before You Move On

1. What causes a shadow?
2. When does light refract?
3. **Infer** You can see a reflection of yourself in a shiny spoon. What can you infer about what the spoon is doing to the light waves?

Heat

Why does popcorn pop? It's all about heat. The popper heats the popcorn seeds. When the material inside the seeds gets hot, it bursts out of the hard outer shell. That's what makes the popping motion. But why does the inner material burst?

The particles that make up matter are always moving. Even the particles in solid materials, such as wood or ice, are moving. The particles have energy because of their motion. The energy of the moving particles is heat energy. The faster the particles of a substance move, the more heat energy they have.

When the particles inside the popcorn seeds are moving very fast, the shell can no longer contain the material. The "explosion" happens, and you get a snack.

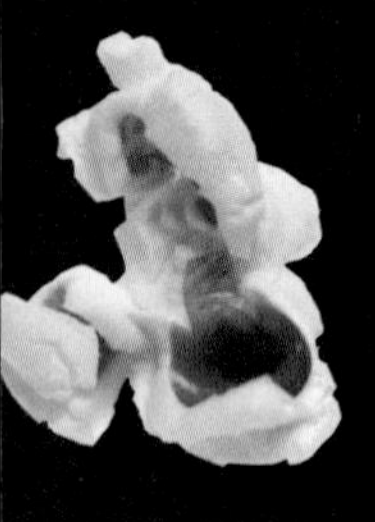

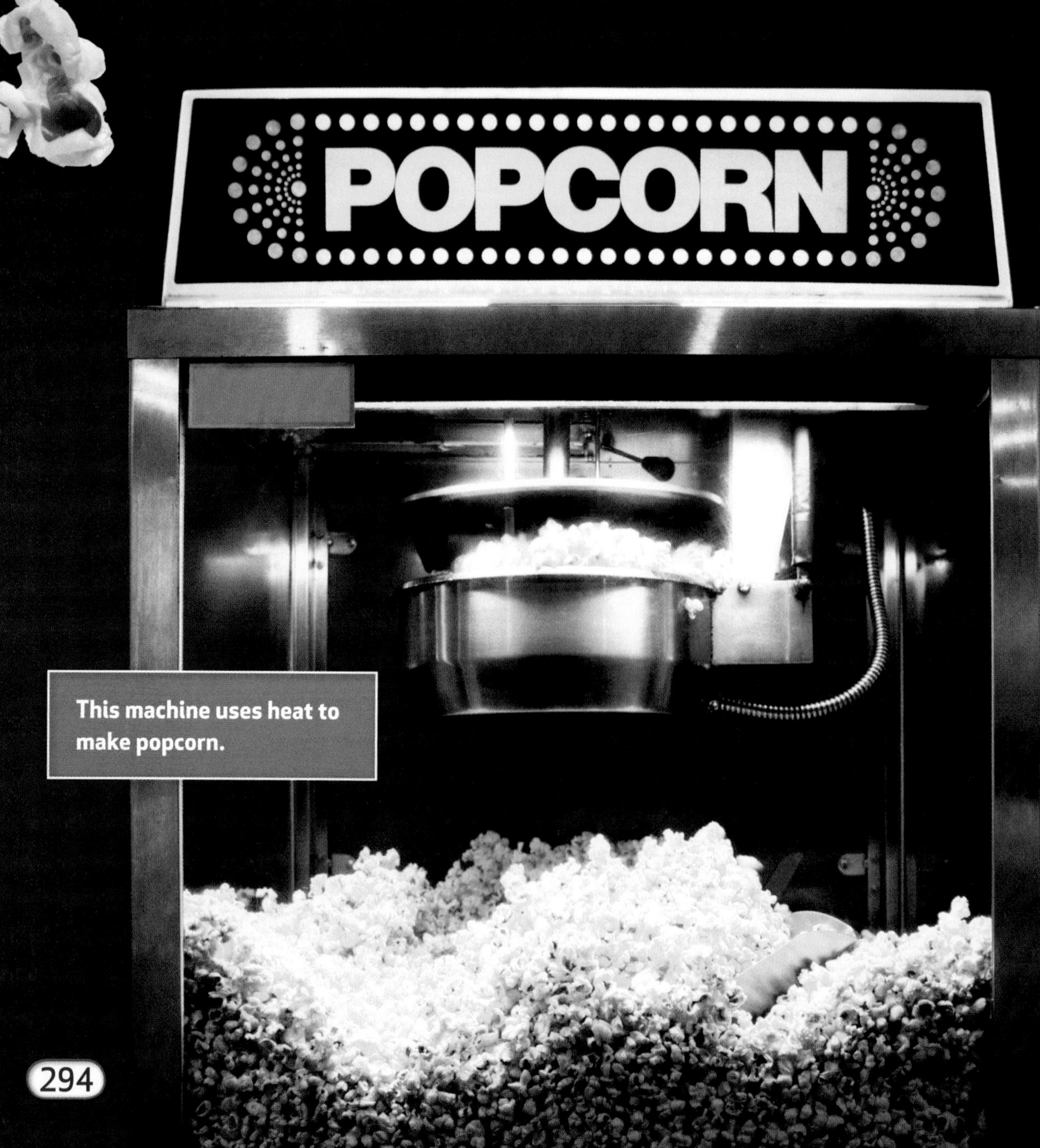

This machine uses heat to make popcorn.

The particles in the pool water are not moving as fast as the particles in the hot popcorn. In this photo, the temperature of the water is being measured. Temperature is the measure of how hot or cold an object is. When particles move slowly, the object or substance has a lower temperature. The swimmers probably hope that the particles in the water are not moving too slowly. They won't want to swim in cold water!

A person uses a thermometer to tell the temperature of the pool water.

Particles in cold substances, such as pool water, move slowly compared to particles in hot substances.

What do you think is happening in the photo? What happens when two objects that have different temperatures touch? Heat energy flows between objects that have different temperatures. You can predict the direction that heat energy will flow.

What happens when you pick up a warm mug? Your hands begin to feel warm. Your hands are cooler than the mug. Heat energy flows from the mug to your hands.

Heat energy flows from your body to the cold air when it is cold outside. Heat energy also moves from a hot cup of cocoa to your hands.

What happens when you pick up a snowball? Your hands begin to feel cold. They feel cold because heat is flowing away from your hands to the snowball.

Heat always flows from warmer objects to cooler objects. The particles in the cooler object start to move faster. The object warms up. The warmer object cools down. Heat flows until the objects eventually reach the same temperature.

Heat flows from warm hands to a cold snowball.

Before You Move On

1. What is heat energy?
2. How does temperature relate to the speed of particles?
3. **Predict** In what direction will heat flow when cold water is added to a hot pan?

Chemical Energy

Your body runs on energy. Every time you take a bite, you are fueling your body. What kind of energy can your body use? The answer is **chemical energy**. Chemical energy is energy that is stored in substances. It is a type of potential energy because it is stored energy.

When you eat, the chemical energy from the food is stored in your body. It changes into other types of energy. Some of the chemical energy becomes heat energy. It keeps your body warm. Chemical energy changes into mechanical energy when you move. Your body also uses the energy in food to build the nutrients you need to grow.

Our bodies use the chemical energy stored in food.

Chemical energy can change into other types of energy. Sometimes burning causes this change. Natural gas is a fuel that contains chemical energy. What type of energy do you observe when a natural gas burner is lit? You see light energy. If you were standing nearby, you would feel heat energy. And, of course, the heat energy heats the pan sitting on the burner. Some of the chemical energy in the natural gas becomes light energy. Some becomes heat energy.

The chemical energy stored in natural gas is released when it burns.

Cars also need fuel to work. Most cars use a fuel called gasoline. Gasoline contains chemical energy. Burning gasoline changes the chemical energy into heat energy and light energy. Some of the heat energy can be used to do work.

In a car engine, heat energy changes into mechanical energy. It makes parts of the engine move. Then, the moving engine parts make the car move. Some of the heat energy is not used to do work. It makes the engine warm.

People fill their gas tanks with fuel at a gas station. The car engines burn the fuel.

When energy changes forms, the total amount of energy stays the same. Think of the car engine. Chemical energy changes to heat energy. Some of the heat energy changes into mechanical energy. But no energy is lost.

Fuel is burned inside a car engine. Some of the chemical energy in the fuel changes into heat energy. Then, some of the heat energy changes into mechanical energy.

Before You Move On

1. Give an example of how energy can cause motion.
2. What happens to the chemical energy in gasoline when it is burned?
3. **Apply** Where does the heat energy and light energy of a candle come from?

NATIONAL GEOGRAPHIC

SEARCHING WITH SOUND IN EGYPT

The city of Alexandria, Egypt, is more than 2,000 years old. It has a rich history, but much of the ancient city was lost. Now people are finding clues to its past in an unexpected place—underwater. Parts of the city were covered by sand and rising seas. Today, underwater archaeologists use sound energy to search the sea near the modern city. They are finding the remains of a great city.

Today, some of modern Alexandria's buildings stand nearly on the shore. Some of ancient Alexandria's buildings lie under the water.

A diver recovers a statue of an ibis, a bird that was sacred to one of the Egyptians' gods.

One of the tools underwater archaeologists use is sonar. Sonar uses reflected sound waves to find objects underwater. The archaeologists use sonar to map the bottom of the sea. They send out sound waves to the seafloor. Then a computer makes a picture of what lies at the bottom using the echoes that return. In Alexandria, the archaeologists used the sonar pictures to find spots that looked like they might be parts of old buildings. Then, they sent divers to carefully uncover valuable pieces of the past.

Using sonar, scientists have identified where ancient Alexandria's buildings were. They are shown in yellow on this map.

A gold ring lies on the seafloor where ancient Alexandria once stood. The artist carved a bird into the beautiful blue-green stone.

Conclusion

Energy comes in many different forms. All forms of energy can do work or cause change. An object may have different forms of energy at the same time. It may transfer its energy to another object. Or, an object's energy may change to a different kind of energy.

Big Idea Energy comes in different forms and can be used to do work or cause change.

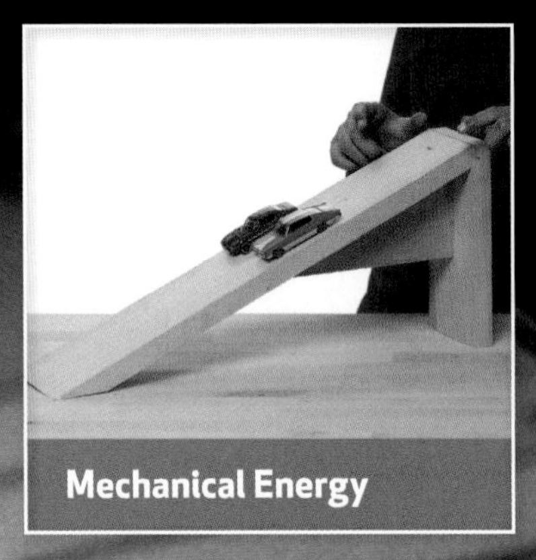
Mechanical Energy

Heat Energy

Chemical Energy

Light Energy

Sound Energy

Vocabulary Review

Match the following terms with the correct definition.

A. **potential energy**
B. **kinetic energy**
C. **mechanical energy**
D. **chemical energy**

1. An object's potential energy plus its kinetic energy
2. Energy that is stored in substances
3. Stored energy
4. The energy of motion

Big Idea Review

1. **Define** What is mechanical energy?
2. **Explain** On what does the kinetic energy of an object depend?
3. **Summarize** What happens when light travels through water?
4. **Predict** Frozen vegetables are added to a pot of boiling water. In what direction will heat energy flow?
5. **Draw Conclusions** Why does a car stop moving when it runs out of gasoline?
6. **Explain** How do bats locate food?

Write About Mechanical Energy

Describe Explain why a tennis ball held above your head has more potential energy than a tennis ball held at your waist.

What Does an Urban Planner Do?

Do you know where the water that flows out of the faucet comes from? How does it get to your house or school? Who figures all of that out? Making sure people have plenty of water is just one part of an urban planner's job. As an urban planner, Thomas Culhane makes decisions about water and energy usage. Urban planners work to make towns and cities nice places to live. They plan where to put schools, parks, and roads. They also help find ways to use energy and other resources wisely and to control pollution.

Thomas Culhane shows a child how to build a solar hot water heater.

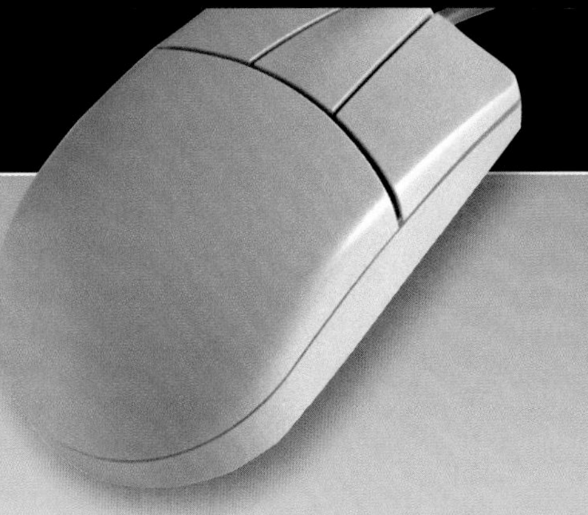

What inspired Culhane? When he was 14 years old, Culhane saw the plans for a model community. It would use renewable resources for its energy sources. The community was never built, but Culhane never forgot the idea.

Urban planners must understand what people need and how they use resources, like energy. Culhane has lived in Guatemala, Lebanon, Egypt, and the United States. Living in different places helps him see how people get food, water, and shelter.

He started Solar CITIES to help people in poor parts of Cairo, Egypt. The people did not have warm water in their homes. Culhane's group helped people build water heaters that used energy from the sun to warm the water.

Urban planners must be creative. They must also be able to solve problems with many different people. Culhane describes his job as being like a hunter-gatherer. He says, "We hunt and gather new knowledge and solutions instead of food."

TECHTREK
myNGconnect.com
Digital Library

Thomas Culhane installing solar panels on a house.

Thomas Culhane uses computers to help him map out how resources will be used.

NATIONAL GEOGRAPHIC

BECOME AN EXPERT

Geothermal Energy: Using Earth's Heat Energy

Most of the energy on Earth originally comes from the sun. Geothermal energy is heat energy from within Earth. It comes from inside our planet, not from the sun. The temperatures deep inside Earth melt rocks and heat water. The temperatures are cooler closer to the surface. Sometimes the melted rocks and hot water make it to Earth's surface. The lava that pours from erupting volcanoes is melted rock. The hot water that reaches the surface forms hot springs and geysers.

Old Faithful, the most famous geyser in Yellowstone National Park, erupts when steam that was heated underground explodes to the surface.

Geothermal energy comes from inside Earth.

Geothermal energy is one of many energy resources people use. Most of the energy people use comes from fossil fuels, such as gasoline and coal. They contain **chemical energy**. Chemical energy is **potential energy** that is stored in substances. People burn them to turn the chemical energy into other types of energy. Fossil fuels are nonrenewable because they form over millions of years inside Earth. Geothermal energy is a renewable resource. You cannot use up the heat energy that is produced inside Earth.

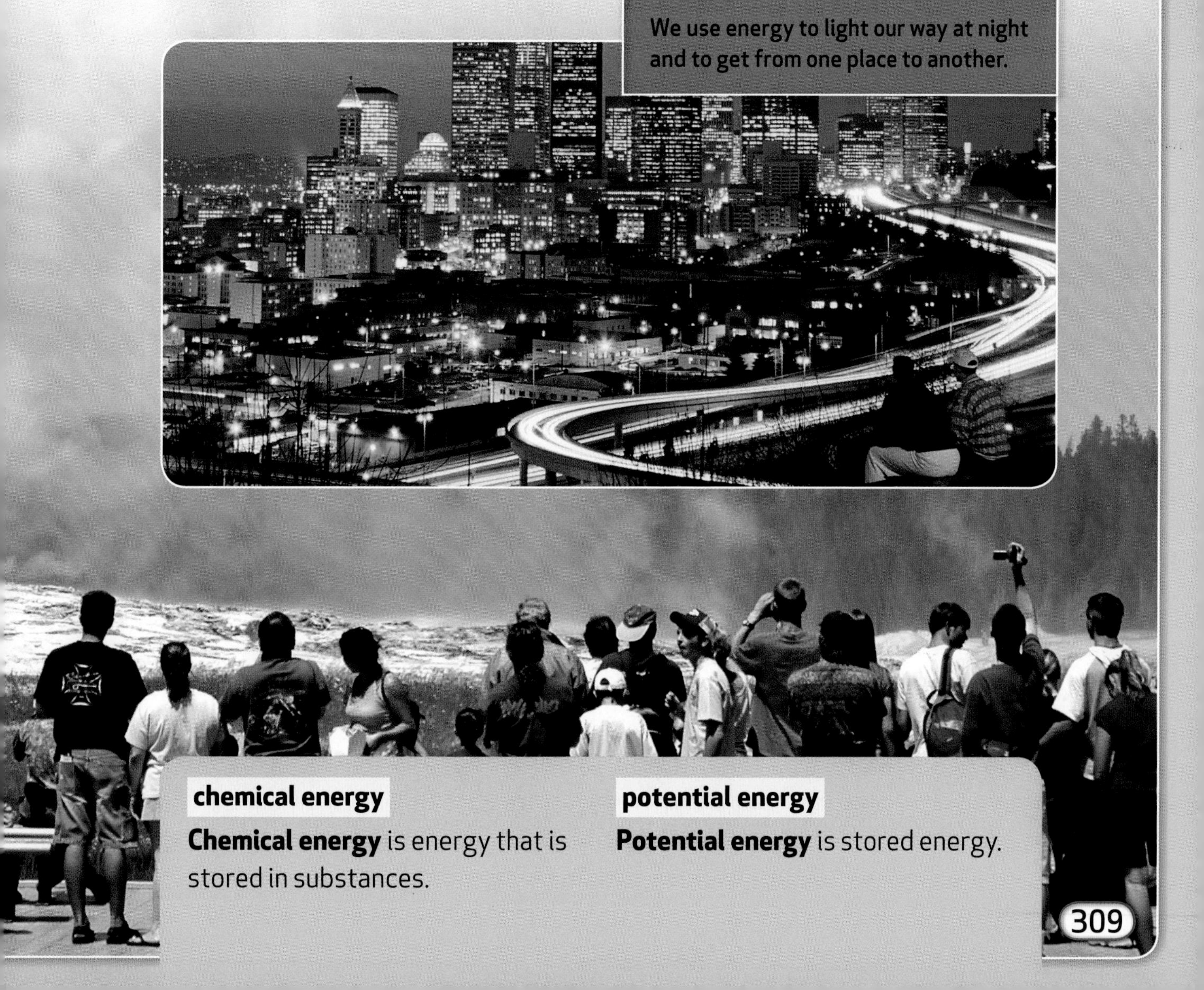

We use energy to light our way at night and to get from one place to another.

chemical energy

Chemical energy is energy that is stored in substances.

potential energy

Potential energy is stored energy.

Geothermal Resources

Most places do not use much geothermal energy. Why? One reason is that different places have different amounts of geothermal energy. In most places, large amounts of geothermal energy are very deep underground. It would be difficult and expensive to reach it.

There are some places where the hot, melted rock is close to Earth's surface. These places have hot springs and volcanoes. California, Nevada, Hawaii, Utah, and other western states have the most geothermal resources. Many countries around the world use geothermal energy, including Iceland, Japan, Mexico, and the Philippines.

U.S.A. GEOTHERMAL RESOURCES

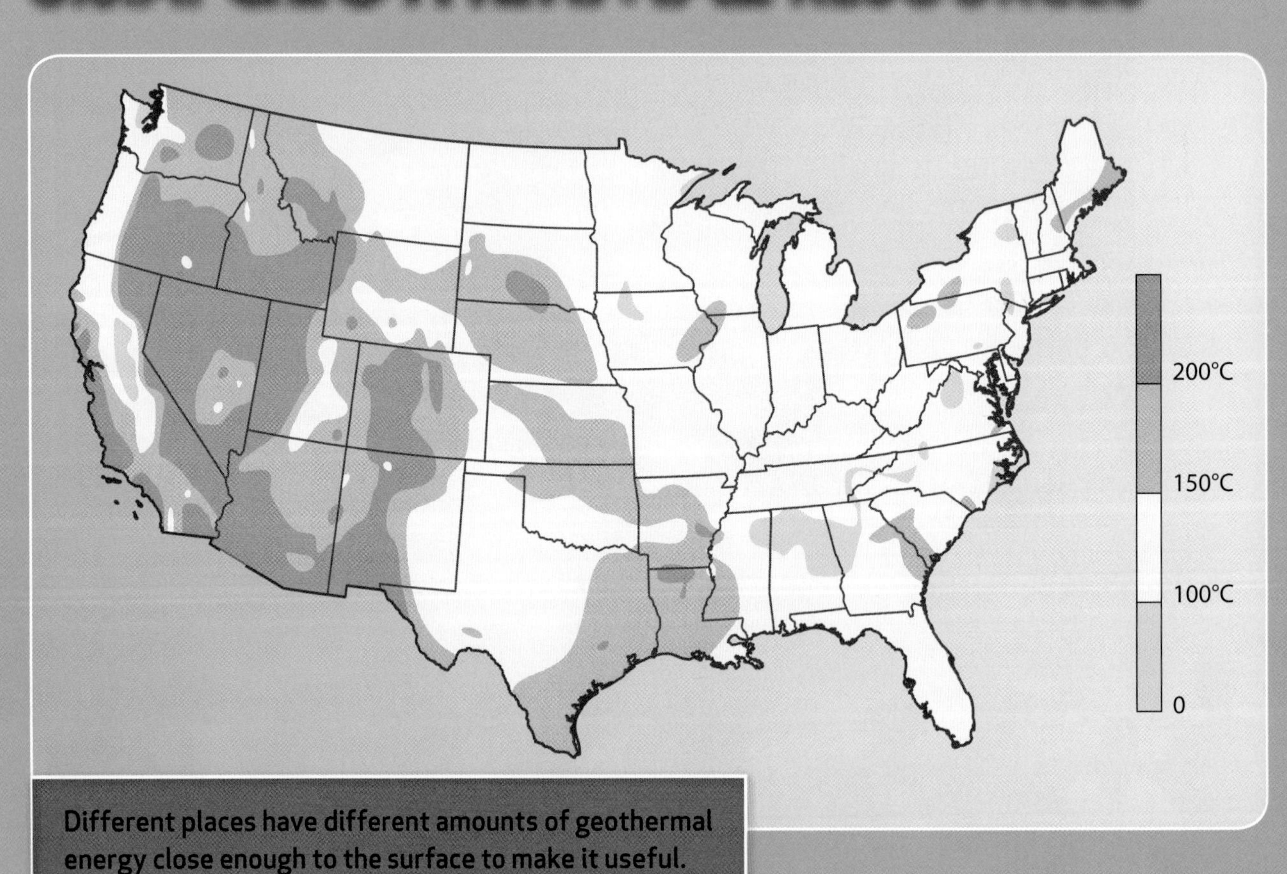

Different places have different amounts of geothermal energy close enough to the surface to make it useful.

People throughout history have used geothermal energy. They used hot springs to cook and bathe. Today, people use geothermal energy in many other ways. The heat energy in the hot water near Earth's surface is used to heat homes, swimming pools, and greenhouses. It is even used to provide warm water for fish farms. The energy in the water is also used to pasteurize milk and to dry onions, garlic, and tomatoes. The direct use of geothermal energy costs less than other ways of heating. It also produces very little pollution.

These shrimp grew up in a special pond heated by geothermal energy.

Workers harvest peppers in a greenhouse heated by geothermal energy.

Heating Buildings Iceland is an island nation that is close to the Arctic Circle. It has more than 20 active volcanoes. It also has plenty of geothermal energy. Most of the buildings in Iceland's capital, Reykjavik, are heated by geothermal energy. The buildings are heated by a district heating system. This system uses hot water that is near Earth's surface. Wells are drilled to reach the water. Then, water that has a temperature of about 80°C is sent through pipes. The pipes go to buildings throughout the city. Heat energy flows out of the hot water. The water becomes cooler. The air in the buildings becomes warmer.

Reykjavik is the world's northern-most capital city. However, warm ocean currents help keep the climate moderate. The average January temperature is -0.3°C (31.5°F). Many places in the United States are much colder than that.

Geothermal heat pumps are used to heat and cool buildings. In the summer, the pipes move heat from the building into the ground. In the winter, the pipes move heat from the ground into the building.

Geothermal heat pumps are very energy efficient and environmentally clean. They can also provide better heating and cooling than many common heating and air conditioning systems.

Geothermal heat pumps provide heat for buildings when it's cold outside. They can also cool buildings in hot weather.

Geothermal Power Plants

Geothermal energy can also produce electricity. Geothermal power plants use steam from hot water underground. In order to reach the water, a deep hole is drilled under ground. The water is then pumped to the surface.

The steam from the water provides the **kinetic energy** needed to power a generator. A generator is a machine that can transform **mechanical energy** into electricity.

HOW DOES A GEOTHERMAL POWER PLANT WORK?

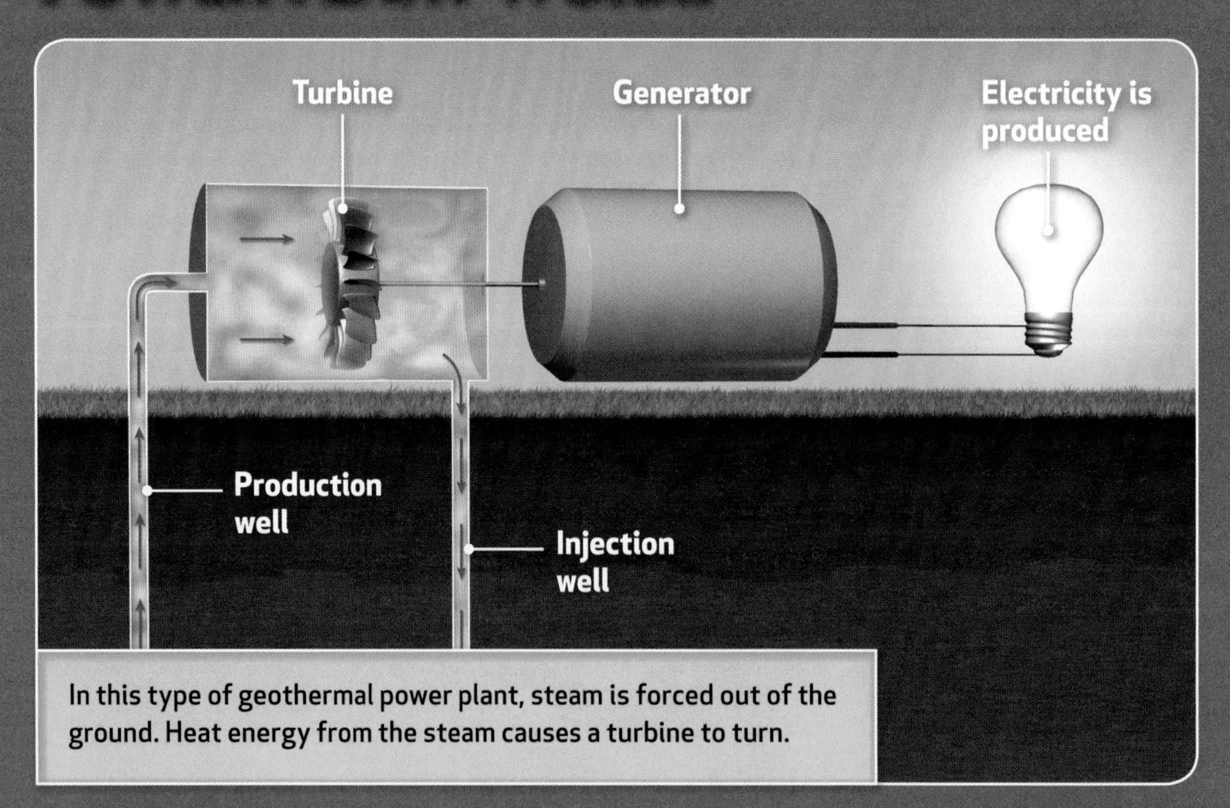

In this type of geothermal power plant, steam is forced out of the ground. Heat energy from the steam causes a turbine to turn.

kinetic energy

Kinetic energy is the energy of motion.

mechanical energy

Mechanical energy is an object's potential energy plus its kinetic energy.

The Geysers power plant in California has been running since 1960.

Earth has a supply of energy that can never be used up. Deep in the crust, water heated by hot rocks is a source of geothermal energy. Different places have different amounts of geothermal energy. People use geothermal energy to heat buildings and to make electricity. It is a renewable and clean source of energy.

TECHTREK
myNGconnect.com
Digital Library

The Blue Lagoon in Iceland offers geothermal seawater for swimmers. The water, which averages 40°C (104°F), is pumped up by the geothermal power plant. That would feel like a very warm bath, no matter what season of the year.

CHAPTER 7 SHARE AND COMPARE

Turn and Talk What are the benefits of using geothermal energy? Form a complete answer to this question together with a partner.

Read Select two pages in this section. Practice reading the pages. Then read them aloud to a partner. Talk about why the pages are interesting.

my SCIENCE notebook

Write Write a conclusion that summarizes what you have learned about energy. In your conclusion, restate what you think is the Big Idea of this section. Share what you wrote with a classmate. Compare your conclusions. Did you recall how energy can come in different forms?

my SCIENCE notebook

Draw Imagine a greenhouse that was heated with geothermal energy. Draw a picture that shows how pipes filled with geothermally-heated water could heat a greenhouse. Add labels to show the direction in which heat would flow. Share your drawing with a classmate. Explain how the geothermal heat in your greenhouse is both the same and different as the geothermal heat in your classmate's greenhouse.

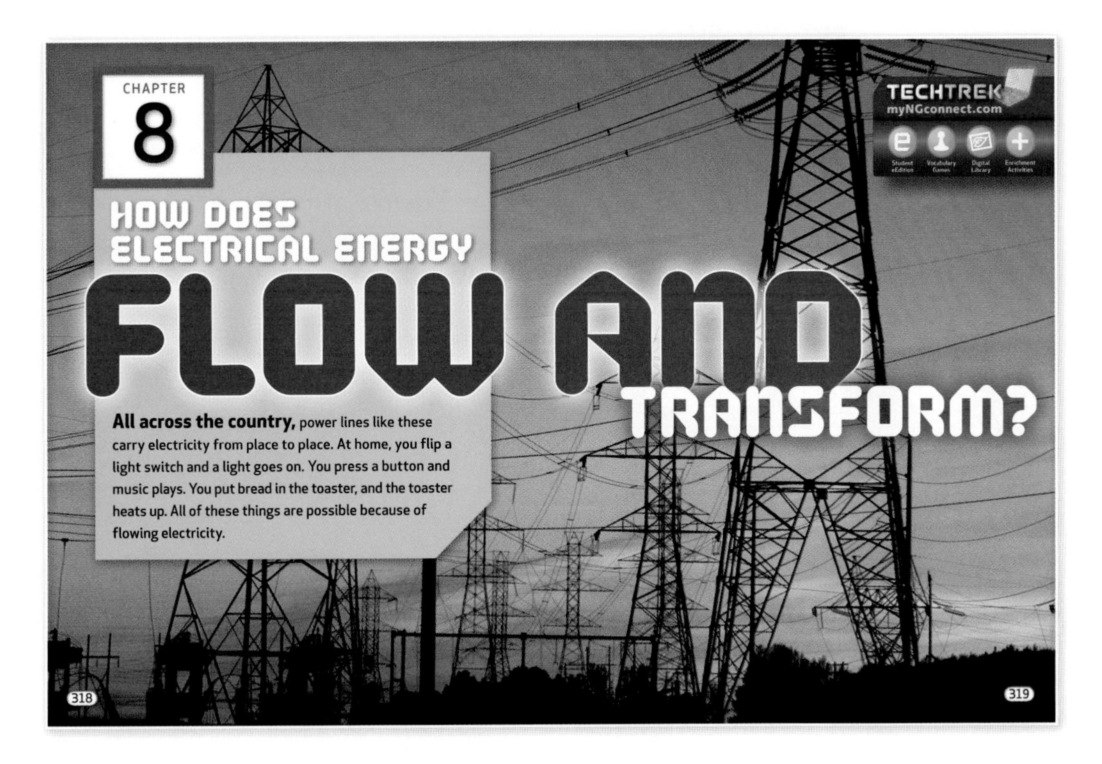

In Chapter 8, you will learn:

FLORIDA NEXT GENERATION SUNSHINE STATE STANDARDS

SC.5.P.10.1 Identify and describe some basic forms of energy, including light, heat, sound, electrical, chemical, and mechanical. **ELECTRICITY, ELECTRICAL ENERGY TRANSFORMS**

SC.5.P.10.3 Investigate and explain that an electrically-charged object can attract an uncharged object and can either attract or repel another charged object without any contact between the objects. **ELECTRICITY**

SC.5.P.10.4 Investigate and explain that electrical energy can be transformed into heat, light, and sound energy, as well as the energy of motion. **ELECTRICAL ENERGY TRANSFORMS**

SC.5.P.11.1 Investigate and illustrate the fact that the flow of electricity requires a closed circuit (a complete loop). **ELECTRICAL CONDUCTORS AND INSULATORS, ELECTRICAL CIRCUITS**

SC.5.P.11.2 Identify and classify materials that conduct electricity and materials that do not. **ELECTRICAL CONDUCTORS AND INSULATORS, ELECTRICAL CIRCUITS**

SC.5.P.11.1 Science in a Snap! Investigate and illustrate the fact that the flow of electricity requires a closed circuit (a complete loop).

CHAPTER 8

HOW DOES ELECTRICAL ENERGY FLOW

All across the country, power lines like these carry electricity from place to place. At home, you flip a light switch and a light goes on. You press a button and music plays. You put bread in the toaster, and the toaster heats up. All of these things are possible because of flowing electricity.

TECHTREK
myNGconnect.com
Student eEdition
Vocabulary Games
Digital Library
Enrichment Activities
AND
TRANSFORM?

SCIENCE VOCABULARY

electricity (ē-lek-TRIS-it-ē)

Electricity is a form of energy that involves the movement of electric charges. (p. 322)

These lanterns need electricity to give off light.

current electricity (KUR-ent ē-lek-TRIS-it-ē)

Current electricity is a form of electricity in which electric charges move from one place to another. (p. 322)

Current electricity carries power from these power lines to the electric appliances in homes.

static electricity (STA-tik ē-lek-TRIS-it-ē)

Static electricity is a form of electricity in which electric charges collect on a surface. (p. 324)

Static electricity attracts the girl's hair to the balloon.

circuit
(SIR-cut)

conductor
(kon-DUK-ter)

current electricity
(KUR-ent ē-lek-TRIS-it-ē)

electricity
(ē-lek-TRIS-it-ē)

insulator
(IN-sū-lā-ter)

static electricity
(STA-tik ē-lek-TRIS-it-ē)

conductor (kon-DUK-ter)

A **conductor** is a material through which electricity can flow easily. (p.326)

Copper is one metal that is a conductor. Electricity flows easily through it.

insulator (IN-sū-lā-ter)

An **insulator** is a material that slows or stops the flow of electricity. (p. 326)

Plastic covers wires that conduct electricity. Plastic is a good insulator.

circuit (SIR-cut)

A **circuit** is a looped path of conductors through which electric current flows. (p.328)

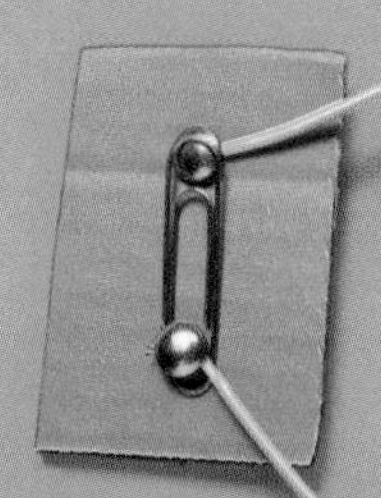

The wires, battery, and light bulb complete a circuit, which is why the bulb is lit.

Electricity

You probably know that to get lights like the ones on this page to light up, you need to plug the cord into an outlet in the wall to get electricity to the lamps. You may know what it's like to sit in the dark when the electricity goes out. But what exactly is electricity? What causes it?

Electricity happens because of electrical charges. Each electrical charge is a tiny bit of energy. All matter is made up particles that have electrical charges. Then why don't you feel electricity every time you touch something? Particles have positive charges and negative charges. In most matter, the number of positive charges is equal to the number of negative charges. That makes the matter neutral. That's why you don't feel any kind of charge when you touch it. Some matter is positive—the matter has more positive charges than negative ones. And some matter is negative—the matter has more negative charges than positive ones. So, what exactly is **electricity**? Electricity is the movement of those charges.

Current Electricity

Most of the electrical machines that you are familiar with use **current electricity**. Current electricity flows along wires. The charges move through the wire so that you can light a lamp, toast bread, or listen to music!

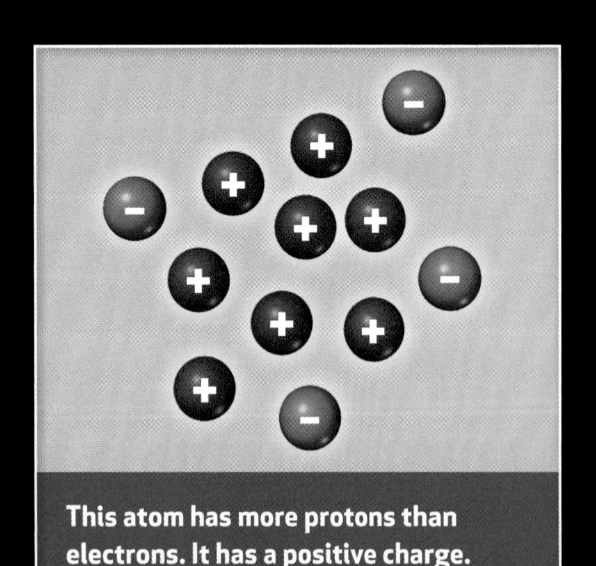

This atom has more protons than electrons. It has a positive charge.

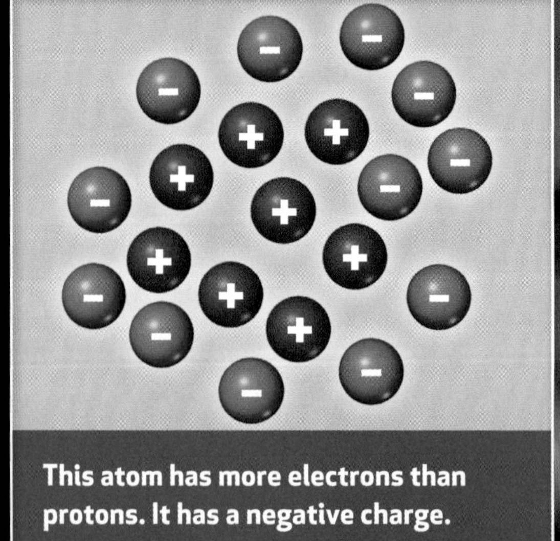

This atom has more electrons than protons. It has a negative charge.

Electricity runs through wires in these lanterns.

Static Electricity When you walk across a carpeted floor, your feet rub on the carpet. Then, when you touch something, such as a person, a pet, or a metal doorknob, you get a shock. What caused the shock? That shock was caused by **static electricity**. Static electricity is a build-up of electrical charges on an object.

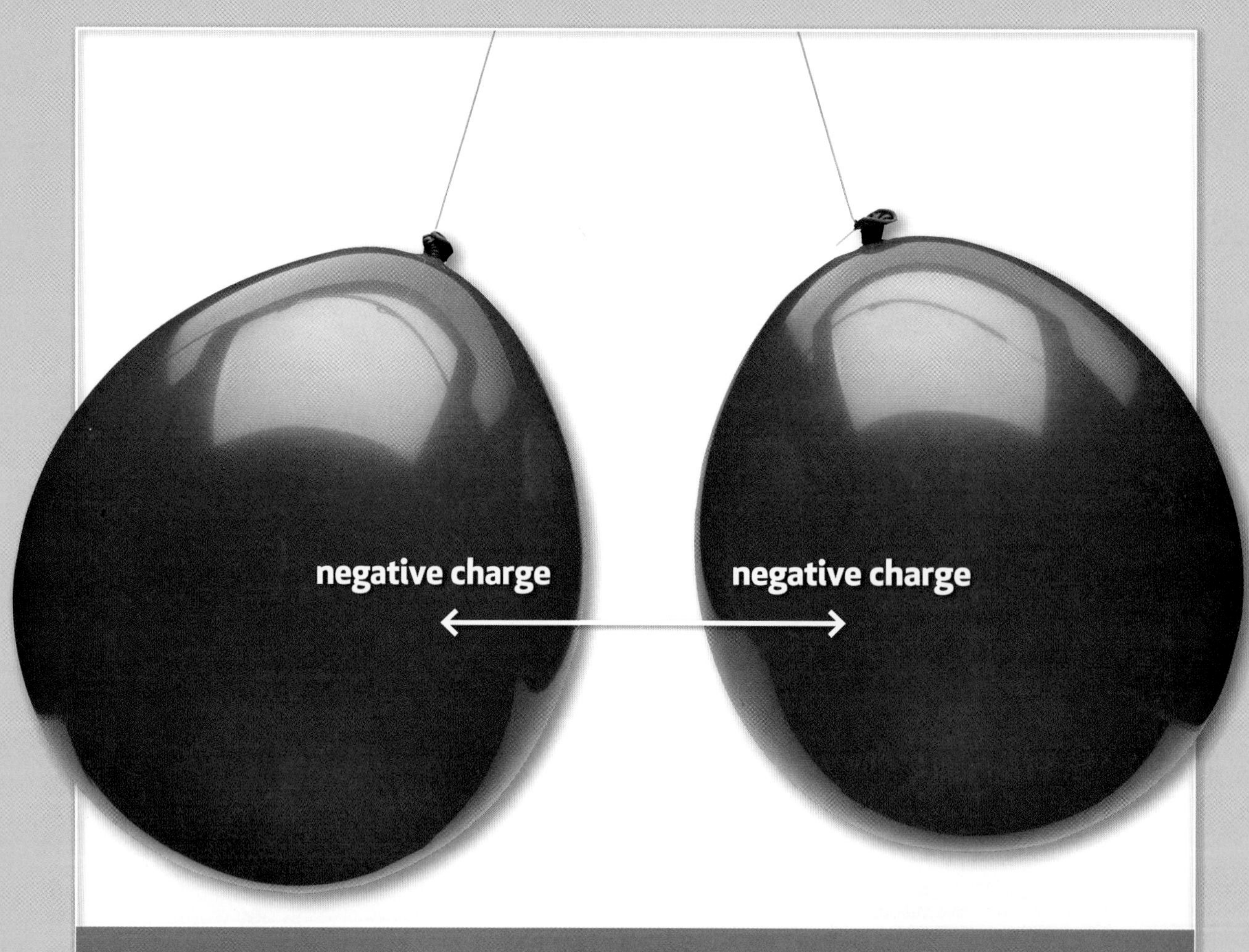

Opposite charges attract. Like charges repel. If you rubbed felt on two balloons, they would both pick up negative charges from the felt. If you brought the balloons close together, they would repel each other, because they both would have a negative charge.

Take a look at the photo below. What can you tell about the balloon and the girl's hair? You know that objects with opposite charges attract each other. If you rub a balloon on your hair, the balloon picks up negative charges from your hair. As those charges build up, static electricity builds, too. If you move that balloon near your hair, the negative changes on the balloon attract the positive charges on your hair. That's what makes hair "stick" to a balloon!

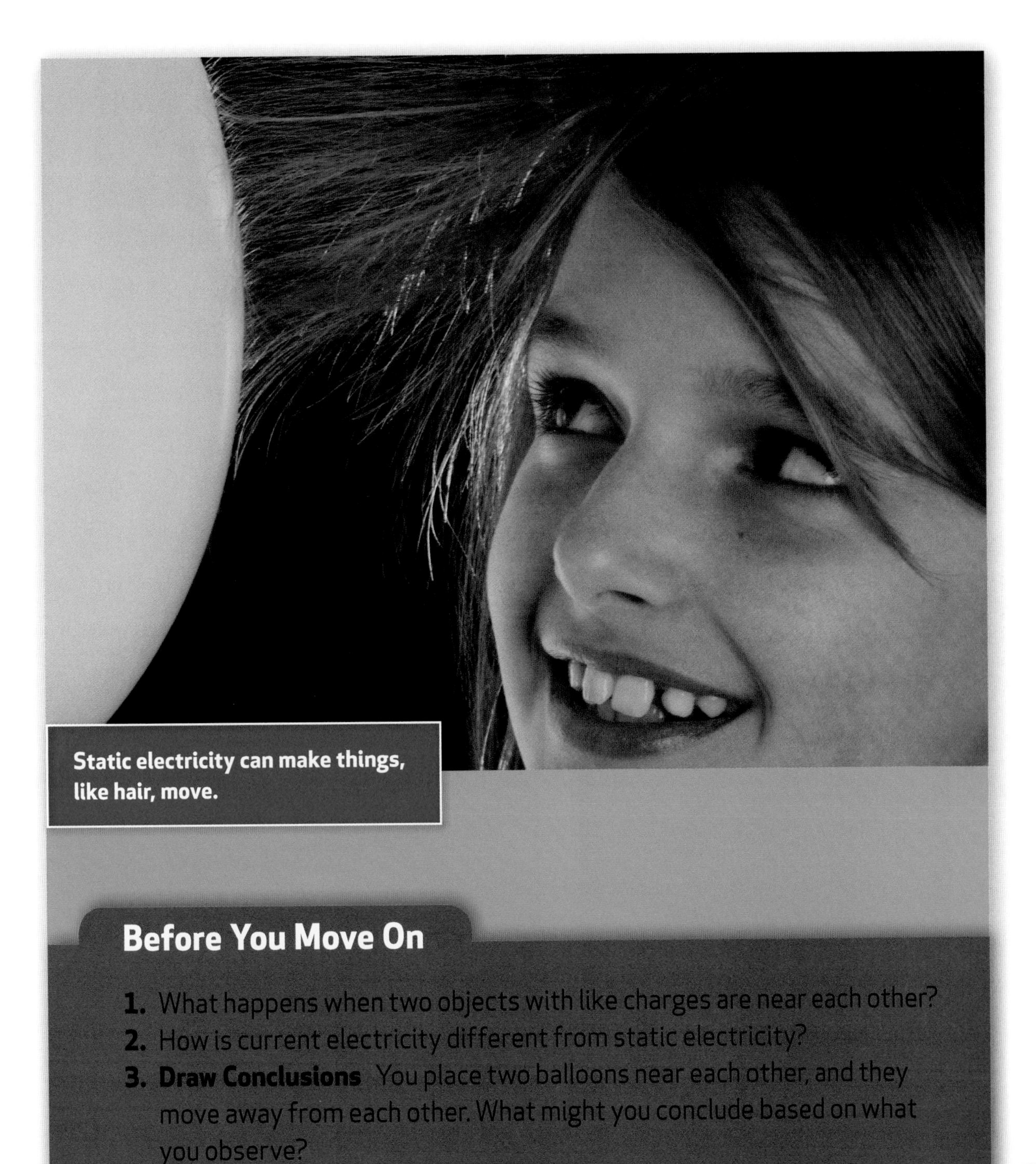

Static electricity can make things, like hair, move.

Before You Move On

1. What happens when two objects with like charges are near each other?
2. How is current electricity different from static electricity?
3. **Draw Conclusions** You place two balloons near each other, and they move away from each other. What might you conclude based on what you observe?

Electrical Conductors and Insulators

Electricity moves from one place to another. Think about a lamp again. Remember how the electricity moves from the wall socket through the wires in the cord to the light bulb? The wires in the cord are conductors. A **conductor** is a material that allows electricity to move easily. Metals such as copper, gold, silver, and iron are conductors. The wires in the lamp cord, for example, are copper.

Have you ever had to get out of an outdoor swimming pool because of lightning? That's because there are materials in water that make water a good conductor. If lightning struck the water, everyone in the pool would be shocked.

Because electricity can be dangerous, it is important to protect people from it. An **insulator** can do this. An insulator is a material that slows or stops the flow of electricity. Plastic, rubber, glass, wood, and ceramics are insulators.

CONDUCTORS	INSULATORS
 Copper wires carry electricity in power lines and electrical plugs.	 Plastic covers electrical wires to keep electricity from shocking you.
 Gold carries electricity in computer motherboards.	 Glass insulators on power poles prevent electricity from reaching a person working on the poles.

Wherever conductors are used, you'll find insulators as well. The gold circuits in a computer are mounted in plastic. Lights and pumps in a pool have plastic, rubber, and glass to separate the current from the water. A telephone is completely covered in plastic and rubber to separate the person from the charge that makes the telephone work. Power lines have ceramic and glass insulators everywhere they connect to a pole.

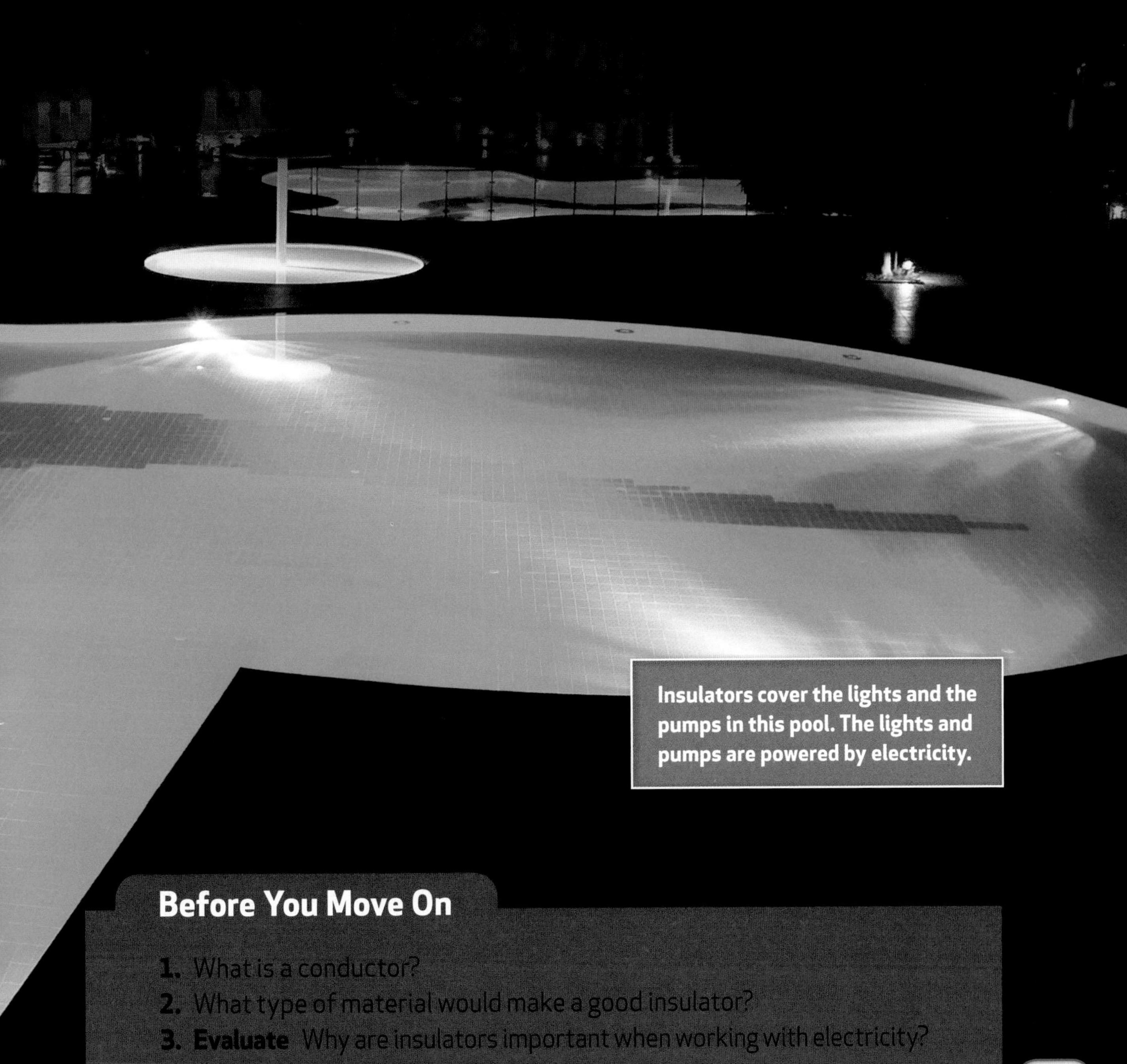

Insulators cover the lights and the pumps in this pool. The lights and pumps are powered by electricity.

Before You Move On

1. What is a conductor?
2. What type of material would make a good insulator?
3. **Evaluate** Why are insulators important when working with electricity?

Electrical Circuits

When you run a complete lap around an oval track, you complete a **circuit**. A circuit is a line or a route that starts and finishes in the same place. Just like a runner on a track, electricity travels in circuits, too. Let's look at a battery to see how this works.

One end of a battery, the one with the bump, is marked positive. The other end, which is flat, is marked negative. If you attach a wire from one end to the other, electricity flows from the negative end to the positive end of the battery. The wires create a circuit—a path through which electricity can flow.

For electricity to flow, a circuit must be closed. Think about that oval track again. If a big hole opened up in the track, you would not be able to complete your circuit. If an electrical circuit is open, the electricity cannot complete its circuit—it cannot flow. So, what happens when you turn a light on? You actually close a circuit, which allows electricity to flow from the source to the light you turned on. When you turn the light off, you open, or break, the circuit so that electricity cannot flow.

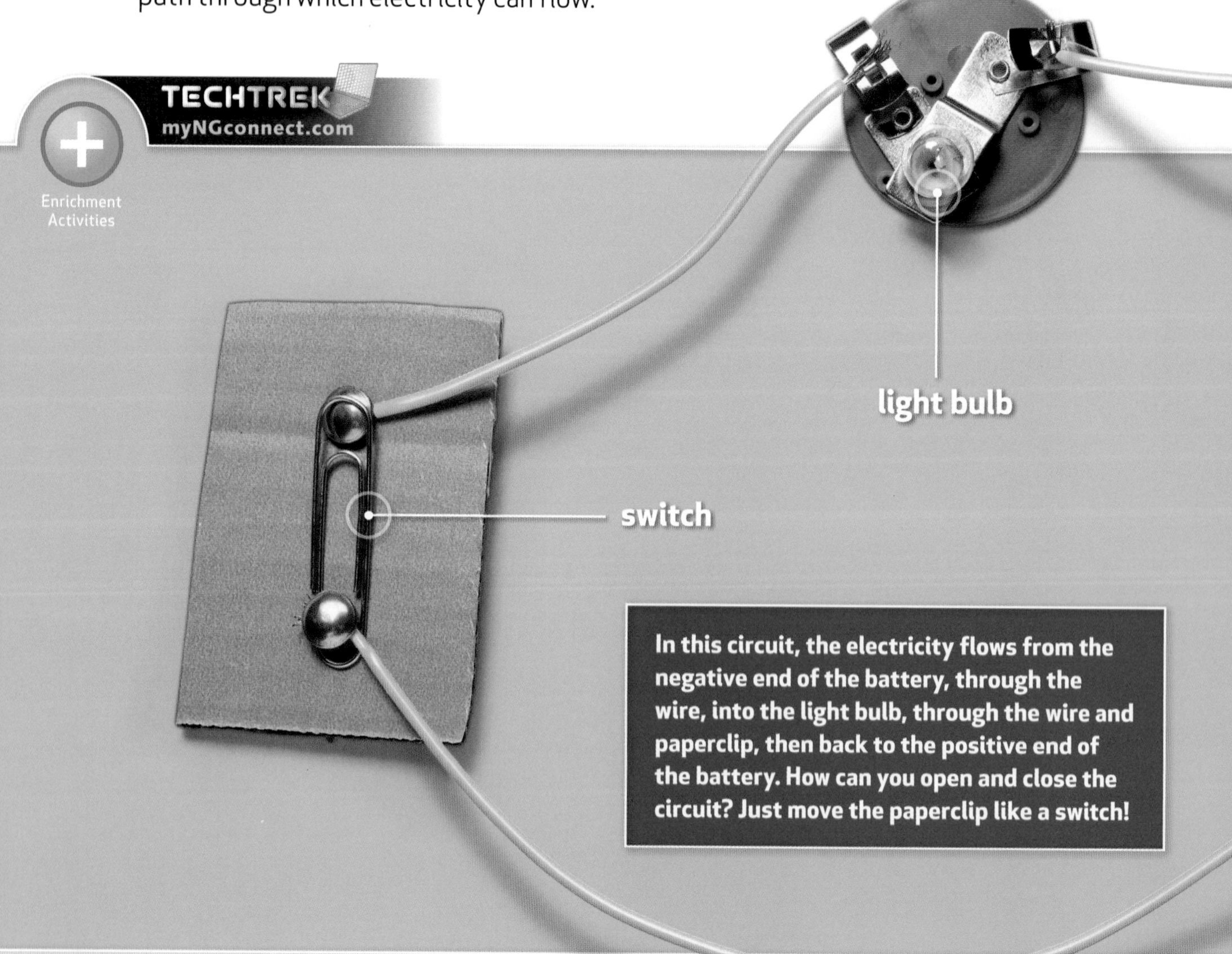

In this circuit, the electricity flows from the negative end of the battery, through the wire, into the light bulb, through the wire and paperclip, then back to the positive end of the battery. How can you open and close the circuit? Just move the paperclip like a switch!

Science in a Snap! Light It Up

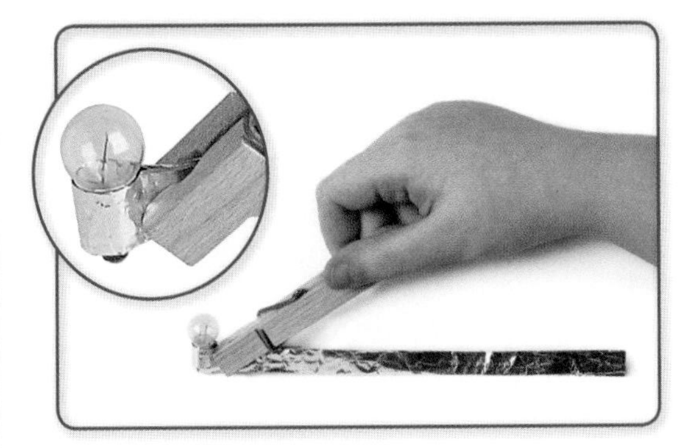

Wrap the foil around the base of the bulb and clip the clothespin over the foil.

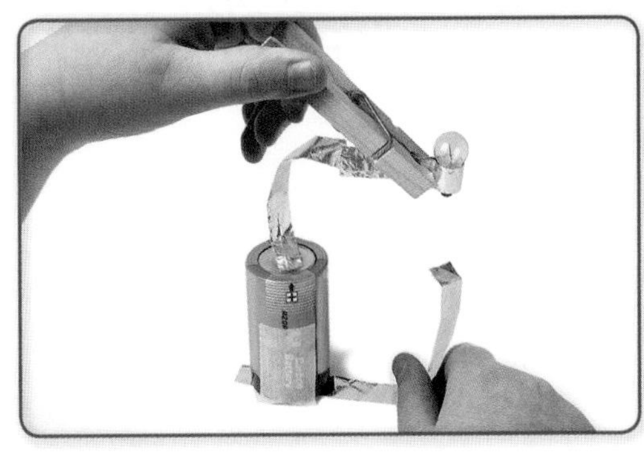

Tape the end of another piece of foil to the battery. Touch the foil strip to the bottom of the bulb to light it. Only touch long enough to light the bulb.

How did electricity travel through your circuit?

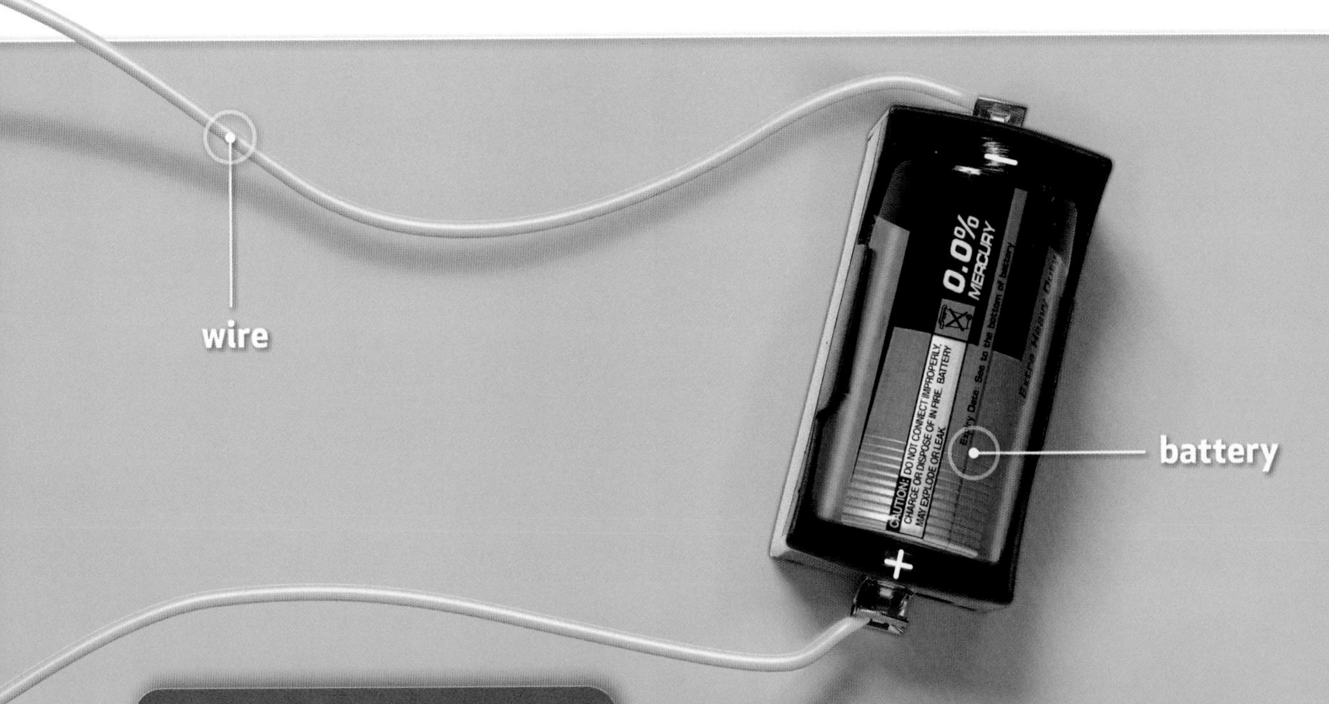

Before You Move On

1. In your own words, tell what a circuit is.
2. What happens to the electricity flowing through the wires in your wall when you flip the light switch?
3. **Infer** How would the circuit inside a flashlight work?

Electrical Energy Transforms

Look at the photograph of Barcelona's Magic Fountain. Lights glow, Water sprays. Music plays in time to the pulses of water. What makes all this happen? Electricity!

Electrical energy can transform, or change, into other kinds of energy. That's what makes electrical energy so useful. If you see light in the room, hear noise from a radio, or feel heat from a heater, you know how useful electrical energy is.

At Barcelona's Magic Fountain, electrical energy is changed into light, sound, and motion.

Take a look at the car below. Why do you think the car is plugged in?

The electric car is run by a battery. The battery provides the electricity. The electricity is converted into motion, heat, light, and sound. The driver plugs the car into a special outlet to recharge the battery.

The battery gives the car enough energy for a driver to warm up the seat, light the road ahead, and play favorite tunes on the radio!

An electric battery in the car creates light, motion, sound, and heat.

Heat At breakfast, you drop a piece of bread into the toaster. You push a button. In a minute or so, your bread is brown and toasted. But what really happened?

Inside the toaster, a special material slowed down the flow of electricity. That caused some wires inside the toaster to heat up. The heat from the wires caused your bread to get brown. You might also have seen the wires glowing. The toaster transformed electrical energy to heat energy and light energy.

When electrical energy moves into the toaster, the electricity changes to heat and a small amount of light.

The chart below shows appliances in the home that change electricity into heat energy. Be careful, though! If these appliances get too hot or you feel heat from a machine that should not get warm, let an adult know. There may be a problem that could cause a burn or even a fire.

COMMON HOUSEHOLD APPLIANCES THAT MAKE HEAT

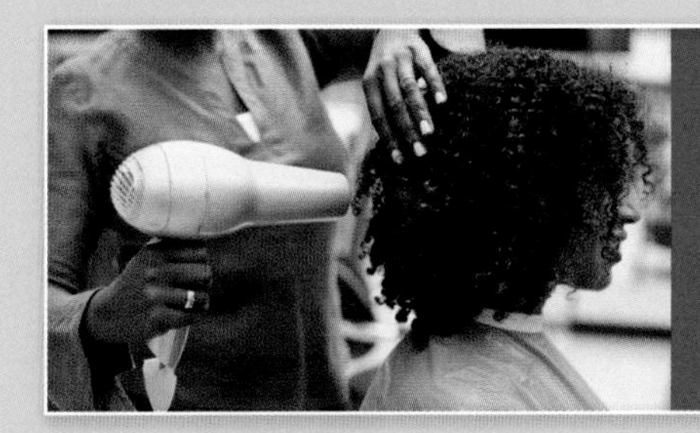

BLOW DRYER

Electrical energy transforms to heat energy and to energy of motion to create heat and blowing air.

OVEN

Electrical energy transforms to heat energy to cook food.

GRIDDLE

A griddle is another appliance that uses heat energy to cook food.

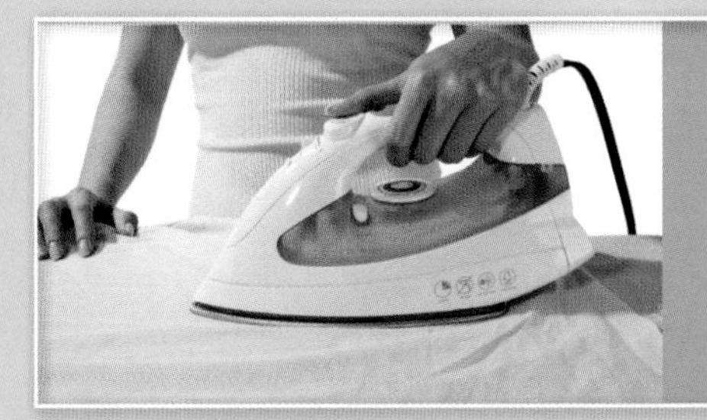

IRON

The heat of the iron smooths wrinkles in your clothing.

Light You can probably name many kinds of lights: ordinary light bulbs, fluorescent bulbs, neon lights, and even lasers. Each of these lights is made differently, but the way each gives off light is the same.

Wires are good conductors. Electrical energy flows along wires to light bulbs. Inside a light bulb, however the electricity slows down. A light bulb has a material inside it called a filament. A filament slows down the flow of the electrical charges and makes the particles shake and forth. As particles shake, they make heat. The shaking particles also make the light bulb glow. That's how electricity is changed to light in a light bulb.

Lights shine all over Hong Kong. Some glow. Some glare. Some look white; some look blue or green. What do they all have in common? Moving electrical charges!

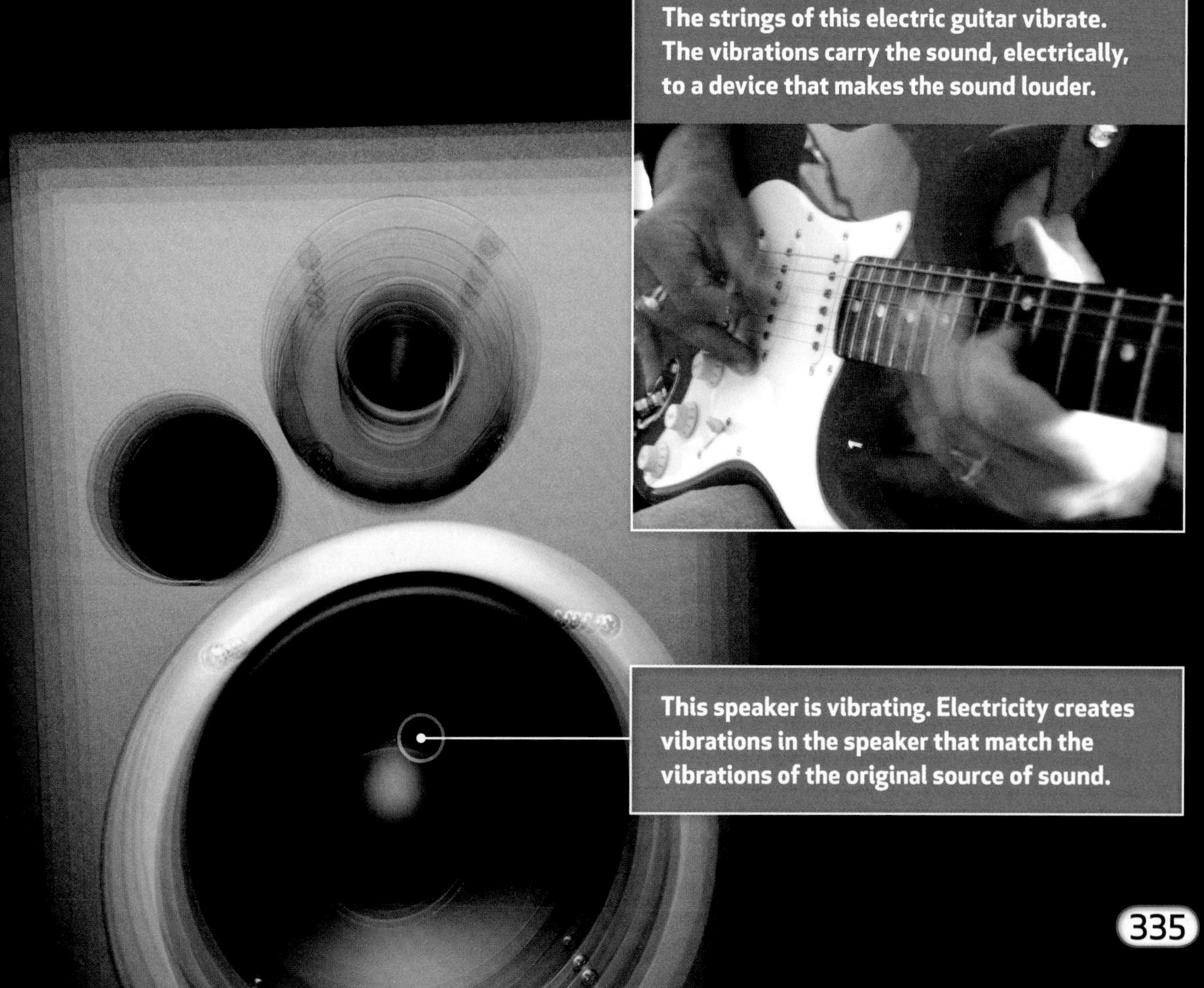

The strings of this electric guitar vibrate. The vibrations carry the sound, electrically, to a device that makes the sound louder.

This speaker is vibrating. Electricity creates vibrations in the speaker that match the vibrations of the original source of sound.

Energy of Motion Many electrical machines and appliances have moving parts. This means that electrical energy transforms into the energy of motion. Electricity works in a special way to create the energy of motion.

Electrical energy transforms into motion and causes the head of this toothbrush to spin.

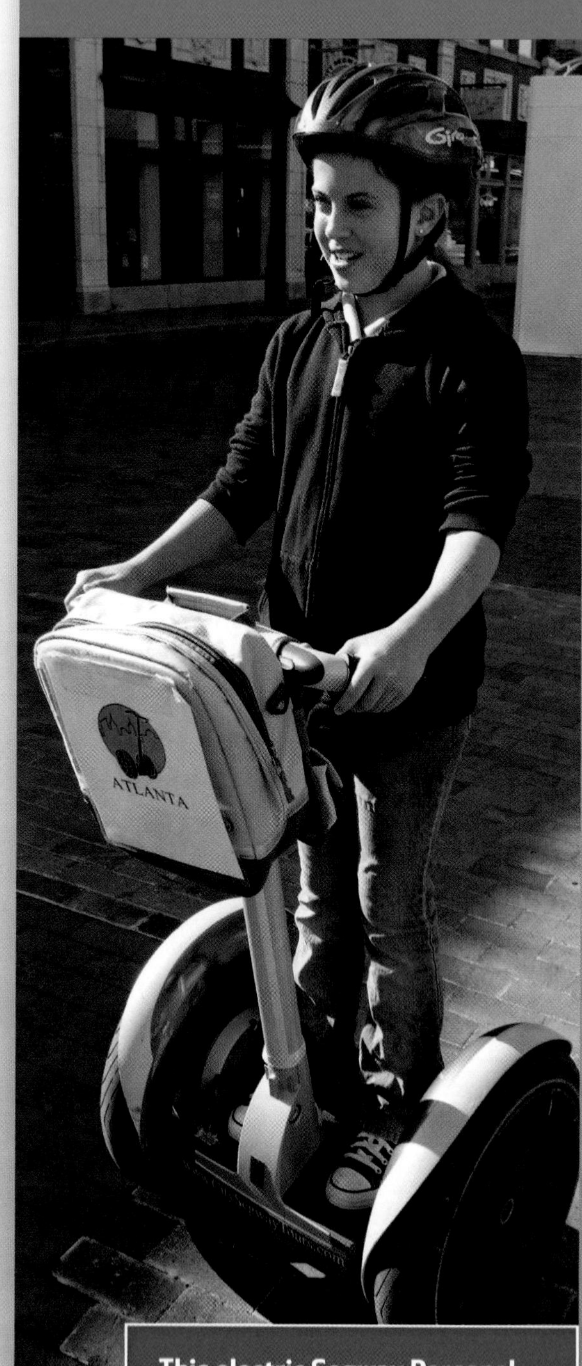

This electric Segway Personal Transporter runs on electricity. The current from a battery runs a motor, which causes the wheels to spin.

Just about anything that changes electricity into motion contains a special part called a motor. A motor has a magnet inside of it. As electricity flows through a motor, the movement of charged particles causes the magnet inside to spin. The spinning magnet makes other parts spin, too. The result of this spinning could be a toy that zips across the floor or a fan with spinning blades.

Look around your home. How many machines with electric motors can you find? A fan above a stove, a garbage disposal, and a turntable in a microwave all use electric motors. Ceiling fans, DVD players, and electric clocks do, too.

Some electric machines need more than one motor. Think about a desktop computer. One motor runs the fan to cool the equipment. Another causes the CD drive to open and close. Electric cars have many motors to control the wheels, the windshield wipers, the fan, sunroof, windows, and more.

An electric motor causes the blades of this fan to move.

Electrical Energy On The Job You already know that electricity can move magnets in a motor. But electricity can also create magnets! A coil with electricity flowing in it creates a magnet.

A magnet created by electric current is called an electromagnet. Electromagnets are temporary. When electricity runs through the electromagnet, the electromagnet attracts some metals. As long as the circuit is closed, the electromagnet continues to attract the metal. When the circuit is opened and the current stops, the electromagnet drops the metal it is carrying.

Making a basic electromagnet is simple. You need a piece of iron, such as a nail, a battery, and some wire not covered in an insulator.

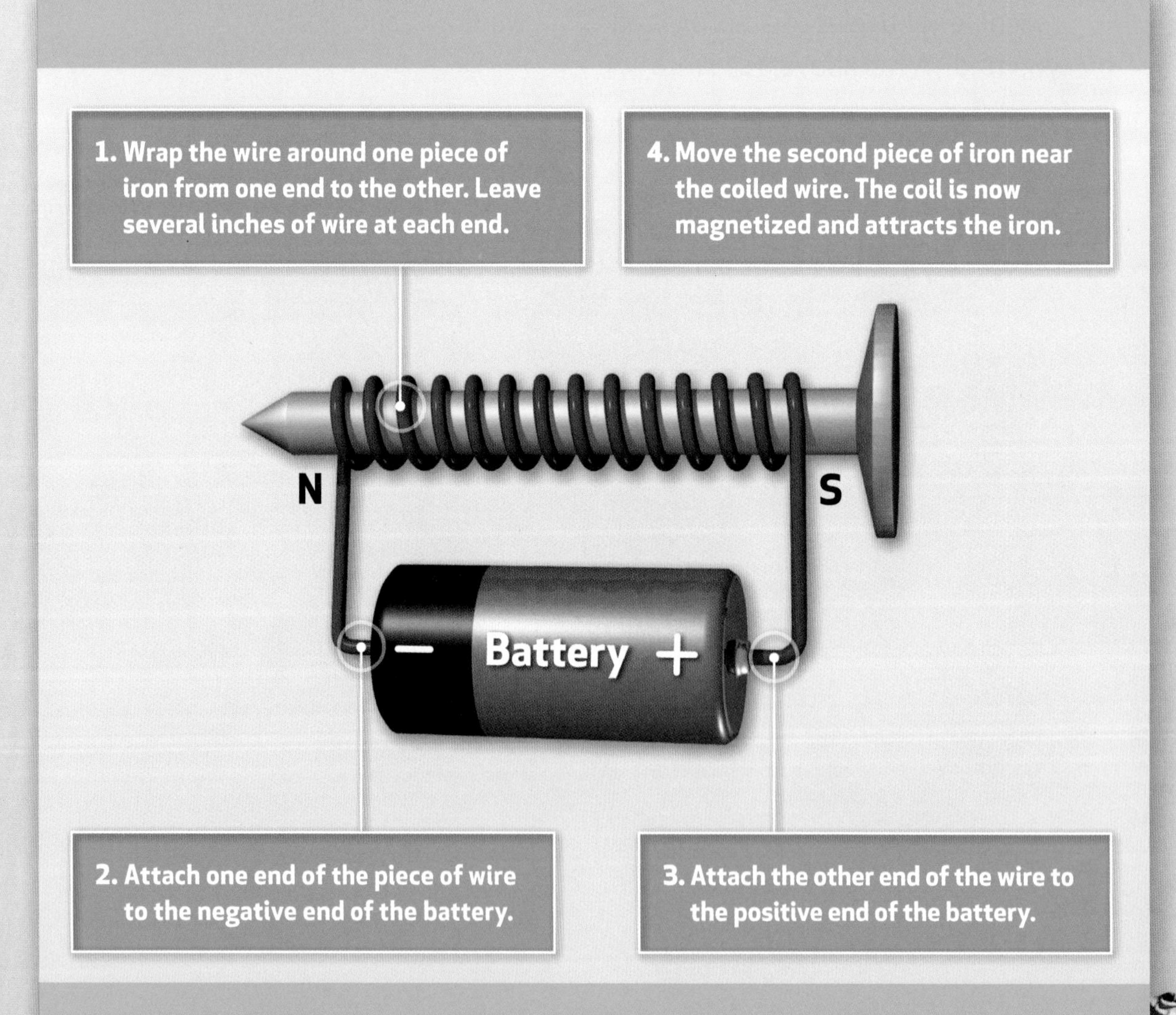

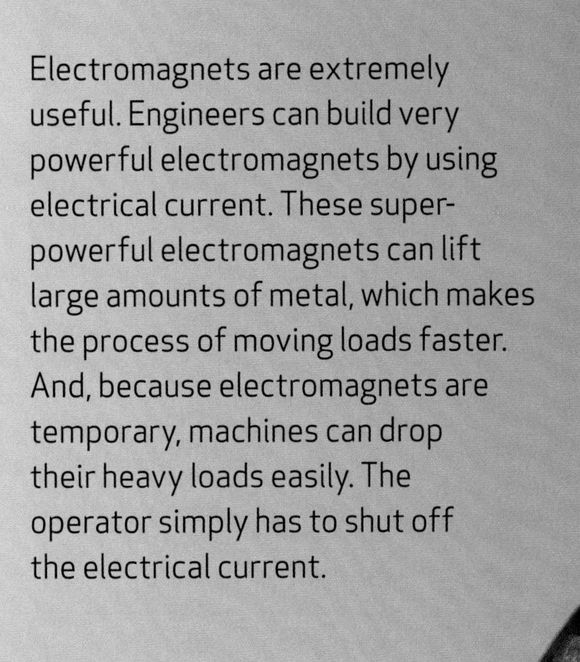

Electromagnets are extremely useful. Engineers can build very powerful electromagnets by using electrical current. These super-powerful electromagnets can lift large amounts of metal, which makes the process of moving loads faster. And, because electromagnets are temporary, machines can drop their heavy loads easily. The operator simply has to shut off the electrical current.

People use electromagnets to move around large amounts of metal.

Before You Move On

1. Electricity transforms into other kinds of energy. What are they?
2. How does electricity create light?
3. **Infer** How can electricity cause a fan to move?

NATIONAL GEOGRAPHIC

WHEN THE ELECTRICITY GOES OUT

On August 15, 2003, a surge of electricity moved through the power lines all over the northern East Coast of the United States. As it did, circuit breakers tripped and cut off electricity to homes and businesses. As the surge moved through the power lines, some power company employees shut electricity down in their towns and cities to keep the surge from reaching them. People call these surges "Blackouts," because the lights go out.

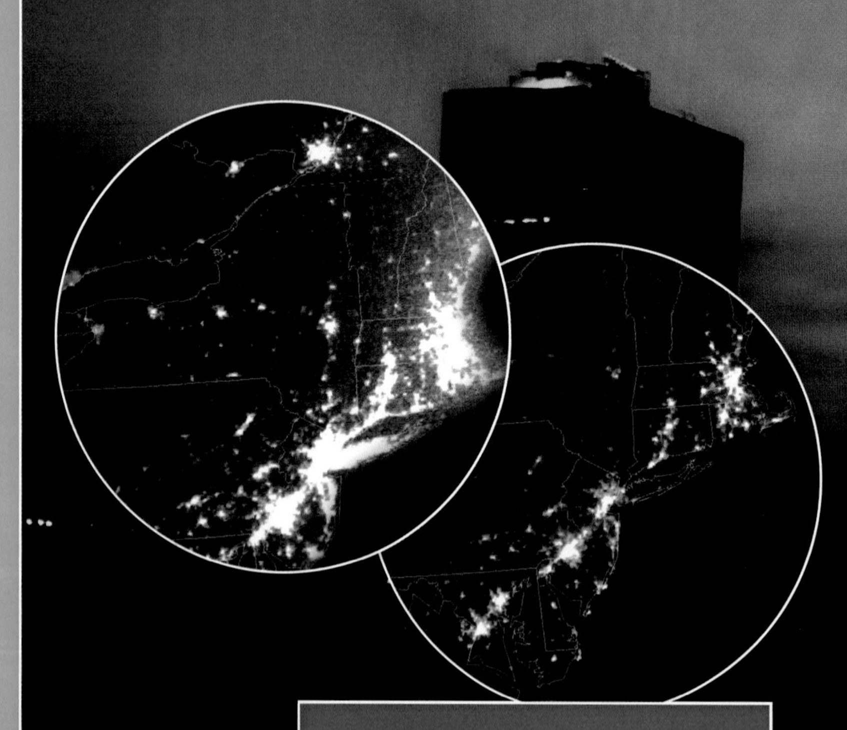

The image on the left shows the lights on the U.S. East Coast that are visible from space. The image on the right shows the same area after the blackout occurred.

A power surge is a sudden increase in the amount of electricity travelling in a current. A small power surge can cause damage to appliances and other electrical machines that are plugged in. A large power surge can damage power plants, power lines, and large electrical machines. In your home, a simple device called a surge protector can protect your appliances from being destroyed by a sudden increase in electricity. But a surge across an electrical system that supplies power to a city or an even larger area is a much bigger problem than a surge in your home. The power surge on August 15th affected more than single homes. When this surge was over, millions of people in the United States and Canada were left in the dark.

People plug their appliances into surge protectors like this one to protect against sudden increases in electricity.

New York City is the largest city in the United States. Power all across the city was shut off in order to protect all of the electrical equipment and power lines. But by turning off the power, everyday life in New York City came to a screeching halt. Thousands of people were stranded underground in dark subway cars. People working in high-rise buildings were trapped without an elevator, and the only way out of the buildings was to go down thousands of stairs in the dark. The people who were on the street had to battle with uncontrolled traffic. All of the traffic lights were off. People all across the city were stranded, stuck in traffic jams, or forced to walk long distances home. The effects of the surge rippled across the United States and Canada, affecting travelers who wanted to fly from New York's airports.

Some power surges come from equipment that does not work properly, or people who use the equipment incorrectly. But power surges can also be caused by nature, such as a lightning strike. Power companies monitor the electricity flowing every minute of every day to try to stop surges before they cause damage.

Times Square, the famous and busy intersection in New York City, is normally flooded with lights.

With no electricity, the only lights in Times Square are from headlights and flashlights.

On August 15, no one was exactly sure what caused the blackout. Officials investigated whether terrorists caused the blackout, but found no evidence that any person had tampered with the electrical grid. Some blamed lightning. The President called for the electrical system to be updated. The blackout could have been caused simply by too many people using electricity to air condition their homes and offices on a hot summer day. Whatever its cause, the blackout showed that people rely on electricity in their everyday lives.

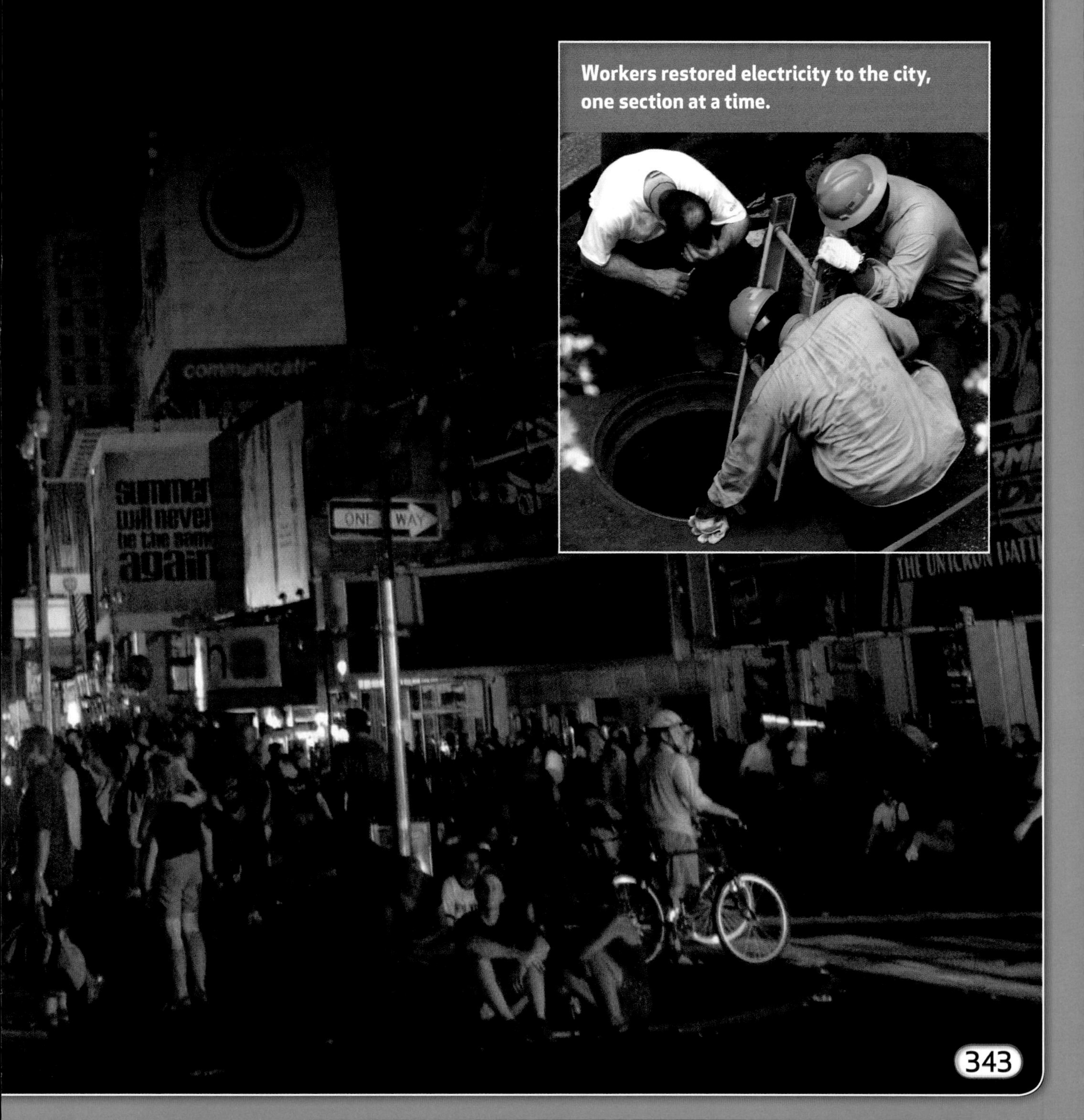

Workers restored electricity to the city, one section at a time.

Electrical charges move to create static and current electricity. Electricity flows easily through conductors, but not through insulators. A closed circuit allows current to flow, while an open circuit stops the flow of electricity. Electrical energy can transform into other forms of energy, such as heat, light, sound, and motion.

Big Idea Electrical energy flows and transforms into other types of energy.

ELECTRICAL ENERGY CAN BE TRANSFORMED INTO . . .

HEAT

LIGHT

SOUND

MOTION

Vocabulary Review

Match the following terms with the correct definition.

A. insulator
B. current electricity
C. conductor
D. electricity
E. circuit
F. static electricity

1. a form of energy that involves the movement of electric charges
2. a form of electricity in which electric charges collect on a surface
3. a form of electricity in which electric charges move from one place to another
4. a material through which electricity can flow easily
5. a material that slows or stops the flow of electricity
6. a looped path of conductors through which electric current flows

Chapter Big Idea Review

1. **Recall** What is electricity?

2. **Classify** Name three materials that are good conductors. Name three materials that are good insulators.

3. **Compare and Contrast** What is the difference between static electricity and current electricity?

4. **Explain** Choose one form of energy: heat, light, sound, or motion. In your own words, explain how electricity changes into that form of energy in a machine you have at home.

5. **Draw Conclusions** If a source of electricity is present, but a light bulb on a circuit is not lit, what can you conclude about the circuit? What would have to happen to make the light bulb go on?

6. **Describe** How would your life change if there was no electricity?

my SCIENCE notebook

Write About Electricity Transforming

Explain What is happening in this photo? How is the electrical energy in the speaker transforming?

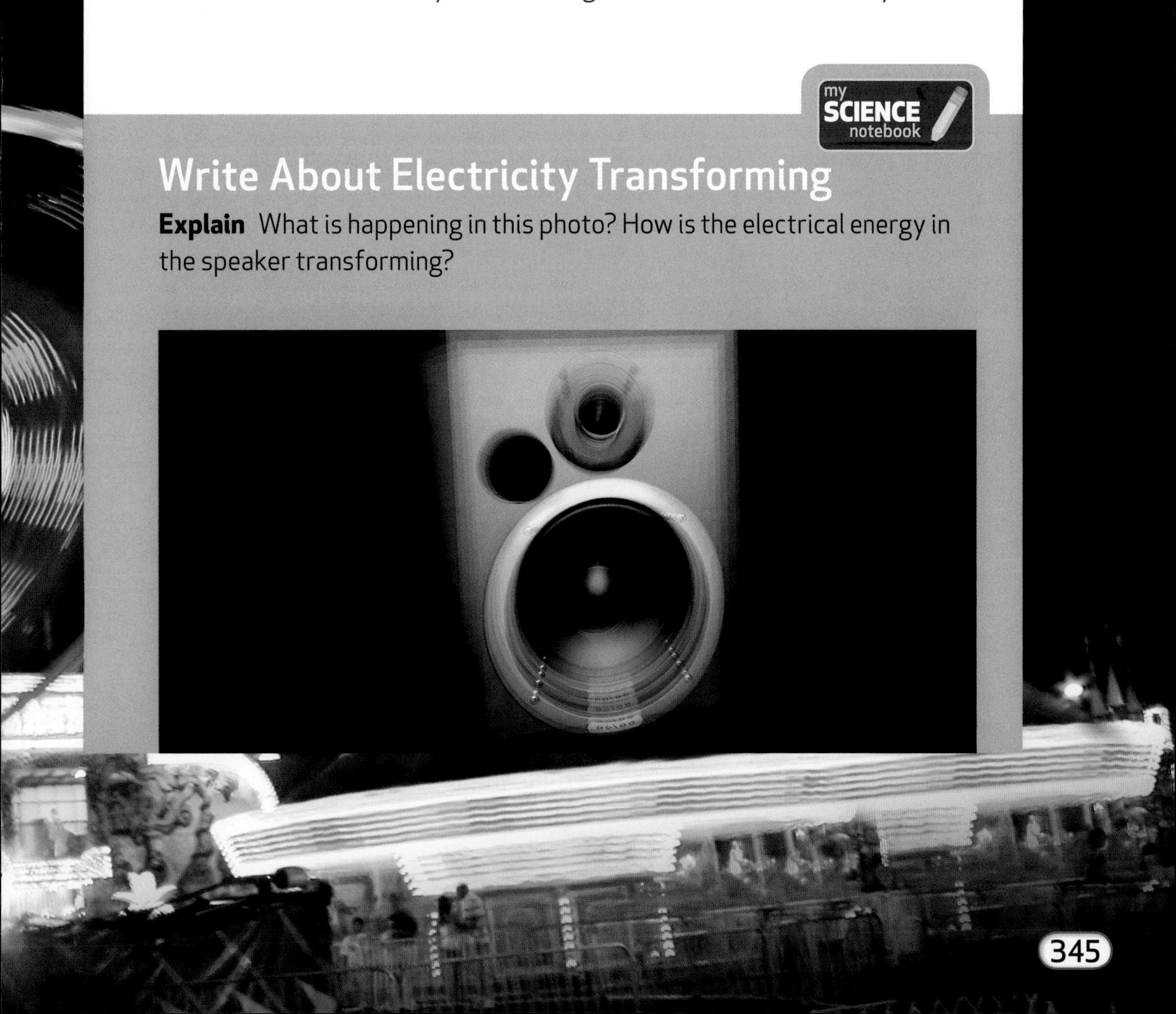

CHAPTER 8

PHYSICAL SCIENCE EXPERT: VIDEO GAME PROGRAMMER

The images, the movement, the sounds—they all have to come from somewhere. Video game programmer Tara Teich puts them together so that video games "come to life" for game players.

Tara Teich

NG Science: What is your job title, and what do you do?

Tara Teich: I'm a video game programmer. I spend my days working at a computer writing the code that brings games to life. My job is to help bring the design, art, and sound together to create gameplay. I help make characters walk when you move the controller and jump when you hit a button. I help make them talk when they see something interesting!

NG Science: What is a typical day at work?

TT: I spend a lot of the day at my desk writing code. But I spend an equal amount of time talking to other engineers. We discuss what they are working on and what is the best way to solve a particular problem. I discuss how the game works with the designers, and I figure out how to make everything more fun.

TECHTREK
myNGconnect.com

Student eEdition

Digital Library

NG Science: What did you have to do to become a video game programmer?

TT: I have a college degree in computer science. This is a degree that would let you be a programmer for any company, not just for a video game company.

NG Science: When did you know that you wanted to be a programmer?

TT: I started programming when I was around 11, and I knew I loved it even then. Being able to bring anything I could imagine to life on the screen was very exciting.

NG Science: What do you like best about your job?

TT: The coolest thing about programming is that there is not just one solution to a problem. I like it that my work is limited only by my imagination. I can write code to do anything that I can think of, if I can puzzle out how to do it.

TECHTREK
myNGconnect.com

Digital Library

Teich uses multiple computers as she programs new games.

NATIONAL GEOGRAPHIC

BECOME AN EXPERT

Video Games: When Electricity Becomes Really Fun

Video games are a huge part of today's culture. Although some parents may doubt their value, kids all around the world use them as a prime source of fun. The thousands of games available today started out as research projects. Professors and students wanted to show what they could get **electricity** to do.

In 1952, a student at Cambridge University created a tic-tac-toe game to run on a university computer. In 1962, students at MIT were able to use the new, smaller computers to create a game called *Spacewar!* It became a hit at colleges all over the country. Never heard of it? That's because the computers needed to play it would take up entire rooms. Only colleges and research facilities had the equipment to play the games.

It took a machine large enough to fill a room to play the simple game of *Spacewar!*

electricity

Electricity is a form of energy that involves the movement of electric charges.

So what do these old computer games have in common with the game systems people use today? All of these games are run by computers.

Computers use **current electricity** to perform instructions. Instructions might be calculations, directions for the cursor to move, or a command to beep. In the video game systems you play today, the directions include making images appear on the screen, sending music and other sounds through the speakers, and sending vibrations to your controller.

Inside the video game system is a computer that works with the television and the controllers to create a fun and exciting game.

current electricity

Current electricity is a form of electricity in which electric charges move from one place to another.

BECOME AN EXPERT

Inside every game system is a microchip. One microchip holds many **circuits**. The circuits are too small to see with the naked eye. The microchip controls what the video game does.

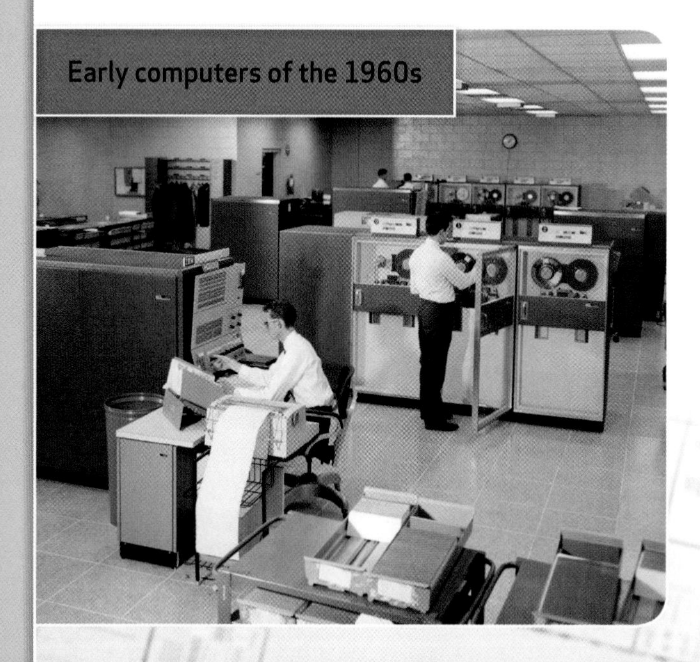

Early computers of the 1960s

If there's an explosion, you'll see it and hear it. And the microchip is also in charge of awarding or taking away points.

The microchip does not know these directions on its own. Each game has code, or instructions, for the game system. This is why you have to change game disks or cartridges to go from one game to the next. Some game systems store some game codes for you. Then, you simply tell the system to load a game. In other words, it goes out to find the correct code.

The microprocessors that game systems use today are much smaller, but far more powerful, than those in the early computers of the 1960s.

circuit

A **circuit** is a looped path of conductors through which electric current flows.

The microprocessor inside any gaming system or computer is attached to a larger set of circuits, called a motherboard. The motherboard is usually made of plastic with copper paths pressed into the surface. The plastic works as an **insulator**. It keeps the electricity that operates the game from reaching people or things it should not reach.

The copper paths on the motherboard create many circuits. These control all of the machinery—the television screen, the controller, and so on. Basically, the motherboard sends out all of the instructions the microprocessor gives. Copper is used on the motherboard because it is an excellent **conductor**.

TECHTREK myNGconnect.com

Digital Library

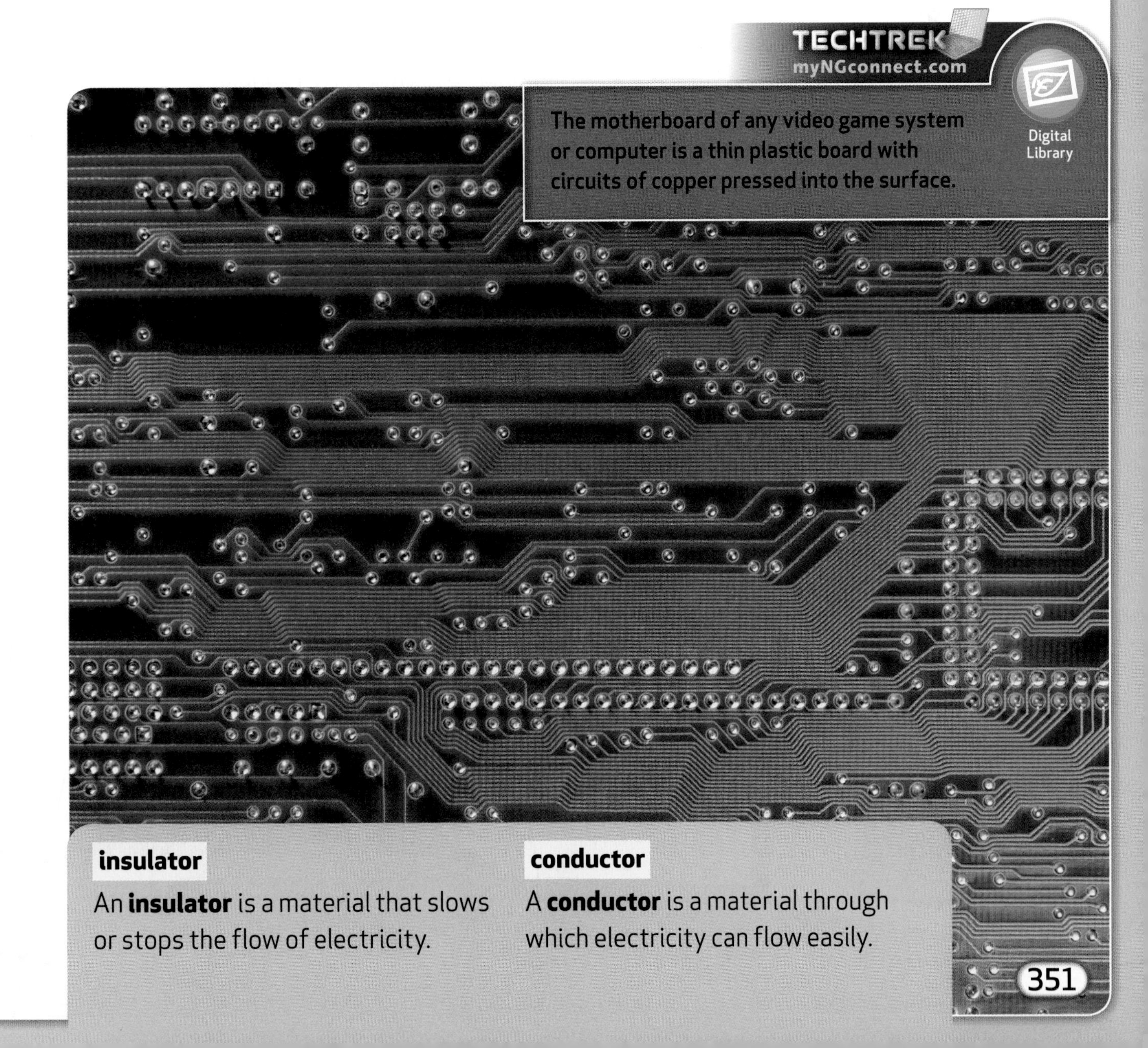

The motherboard of any video game system or computer is a thin plastic board with circuits of copper pressed into the surface.

insulator

An **insulator** is a material that slows or stops the flow of electricity.

conductor

A **conductor** is a material through which electricity can flow easily.

The difference in the systems is the type of computer equipment inside. Each type of game system reads directions in a particular way. It is as if Game A speaks Spanish and Game B speaks Swahili. They just don't understand each other's directions. This explains why you cannot put one game into another game's play system and expect it to work.

This means that if the same video game is available for two game systems, then two entirely different sets of instructions had to be written. The programmers writing the code must understand how both game systems receive and send instructions. The time it takes to write the code can cost the company a great deal of money. This is why many games are available for only one system.

Video games are fun. Believe it or not, they are also useful! People who play games want great graphics. The more things a video game system can do, the better. Many improvements that people have made to computers happened with video games first. People who work with "regular" computers borrow technology from the gaming technology. The same graphics cards that make games fun can work for business, too.

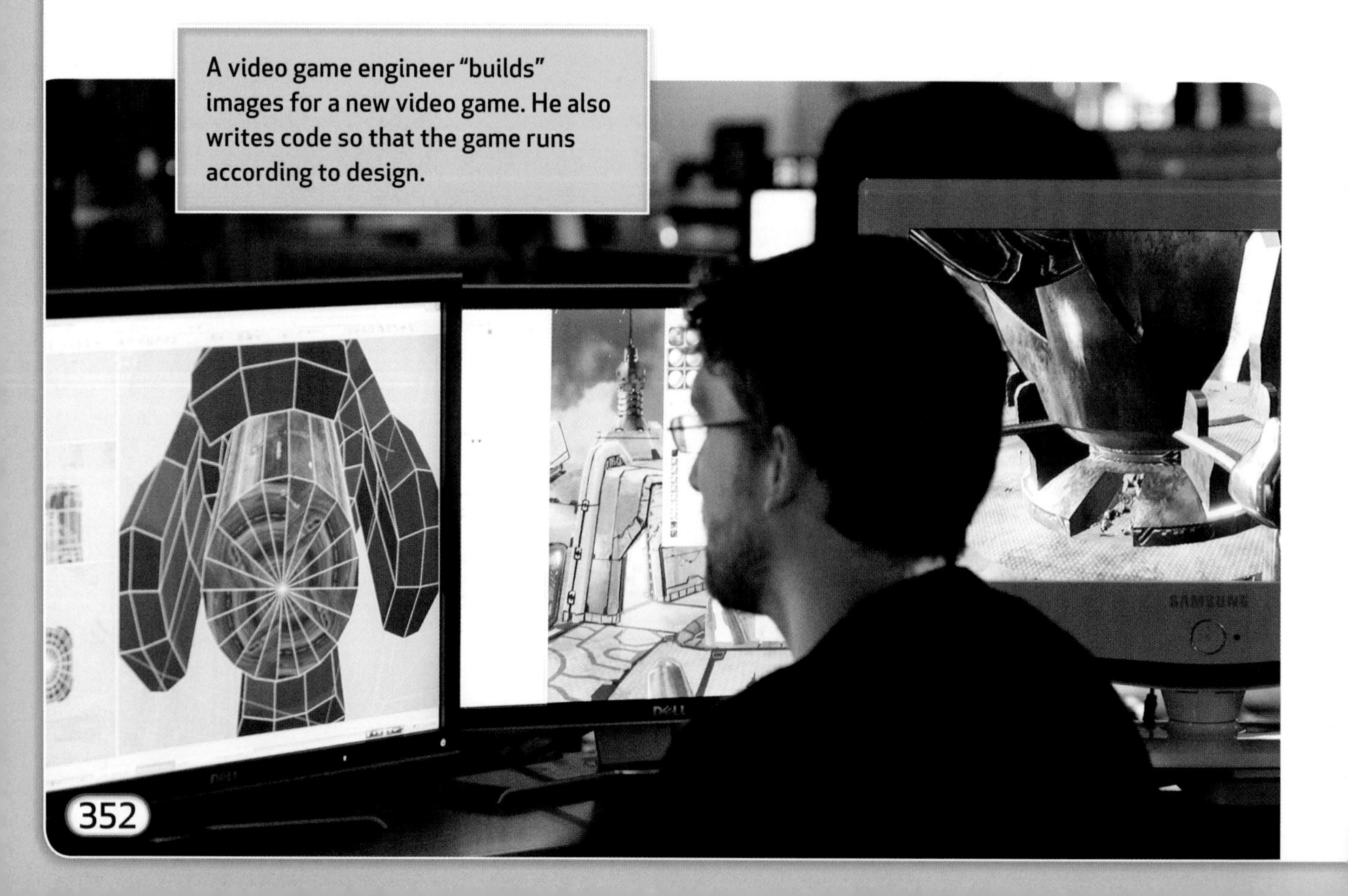

A video game engineer "builds" images for a new video game. He also writes code so that the game runs according to design.

The games themselves can help people on the job. The U.S. military uses video games to teach spies how to think and react. Doctors use video games to help patients getting treatment for cancer. NASA has even developed technology that helps kids with attention deficit disorder (ADD) use video games to improve their focus.

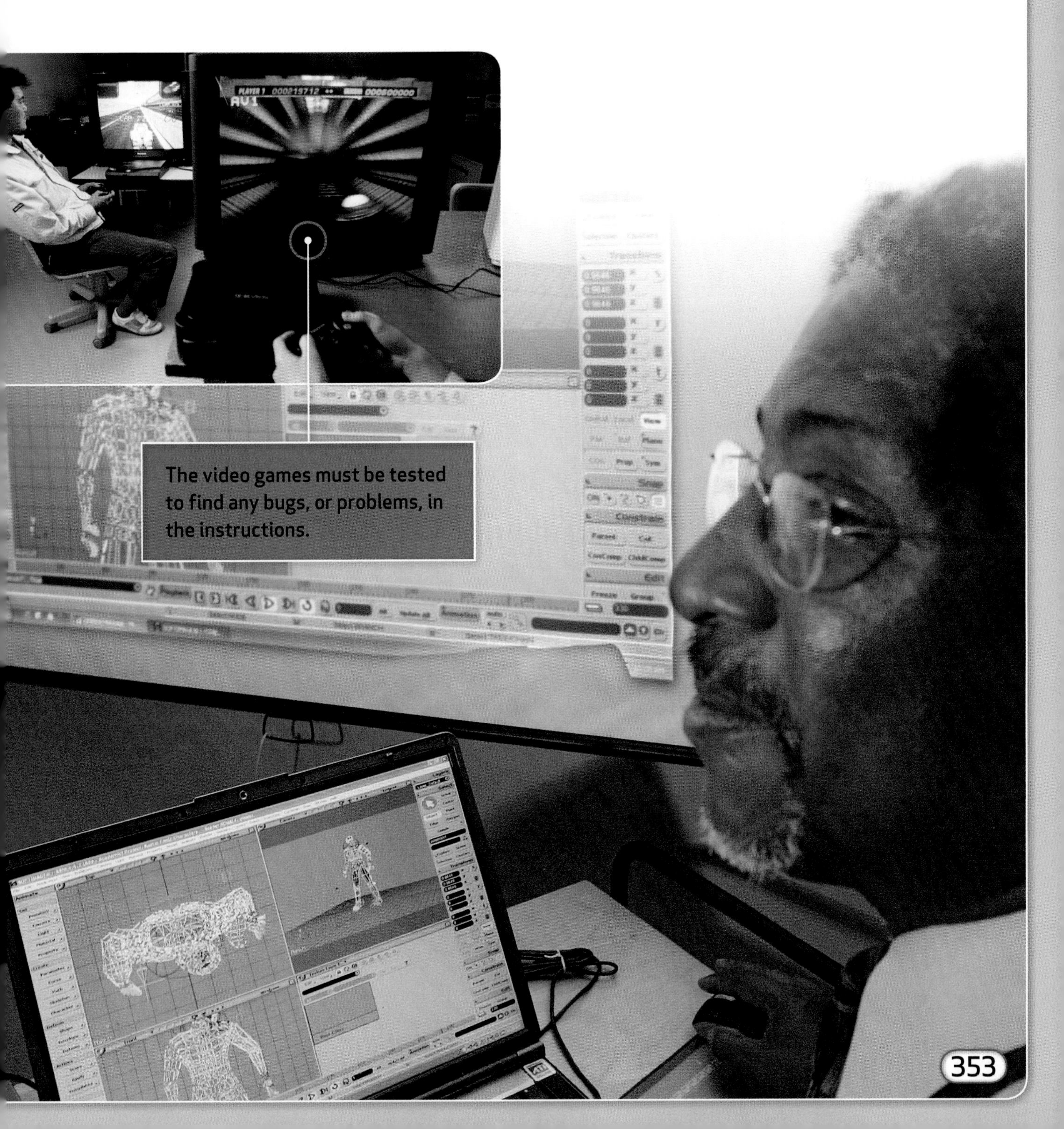

The video games must be tested to find any bugs, or problems, in the instructions.

It's interesting to know about the workings of video game systems. However, that does not mean you should take apart any of the pieces of your game systems.

Do not open up your game systems if you want them to work. As interesting as it is to look inside and become more of an expert, you would very likely cause a problem. You run the risk of passing **static electricity** on to the motherboard or other parts of the computer. This static electricity can damage the circuits and cause the game not to work.

Why is it so important to take care of game disks and cartridges? Dust and scratches can damage the code. The computer will no longer understand it. Then the game won't work.

Taking care of your games will help them last much longer.

static electricity

Static electricity is a form of electricity in which electric charges collect on a surface.

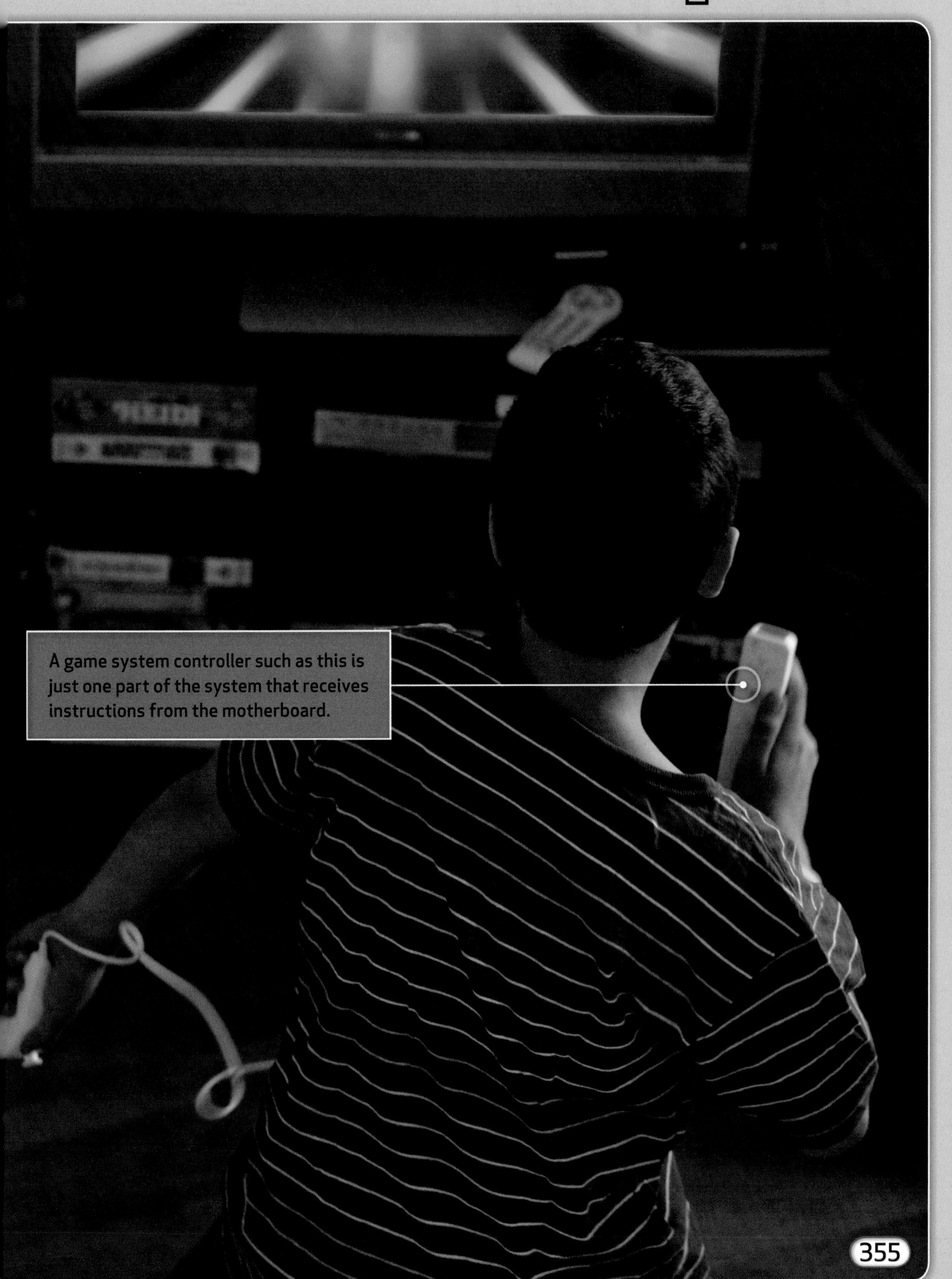

A game system controller such as this is just one part of the system that receives instructions from the motherboard.

CHAPTER 8 SHARE AND COMPARE

Turn and Talk How does electrical energy change into other forms of energy in video games? Form a complete answer to this question together with a partner.

Read Select two pages in this section. Practice reading the pages. Then read them aloud to a partner. Talk about why the pages are interesting.

Write Write a conclusion that tells the important ideas you learned about video games. State what you think is the Big Idea of this section. Share what you wrote with a classmate. Compare your conclusions. Did your classmate make the connection between video games and electrical energy?

Draw Someone asks you, "How does a video game use electrical energy?" Draw a diagram to answer the question. Combine your drawing with those of your classmates to make a video game "How Does It Work?" guide.

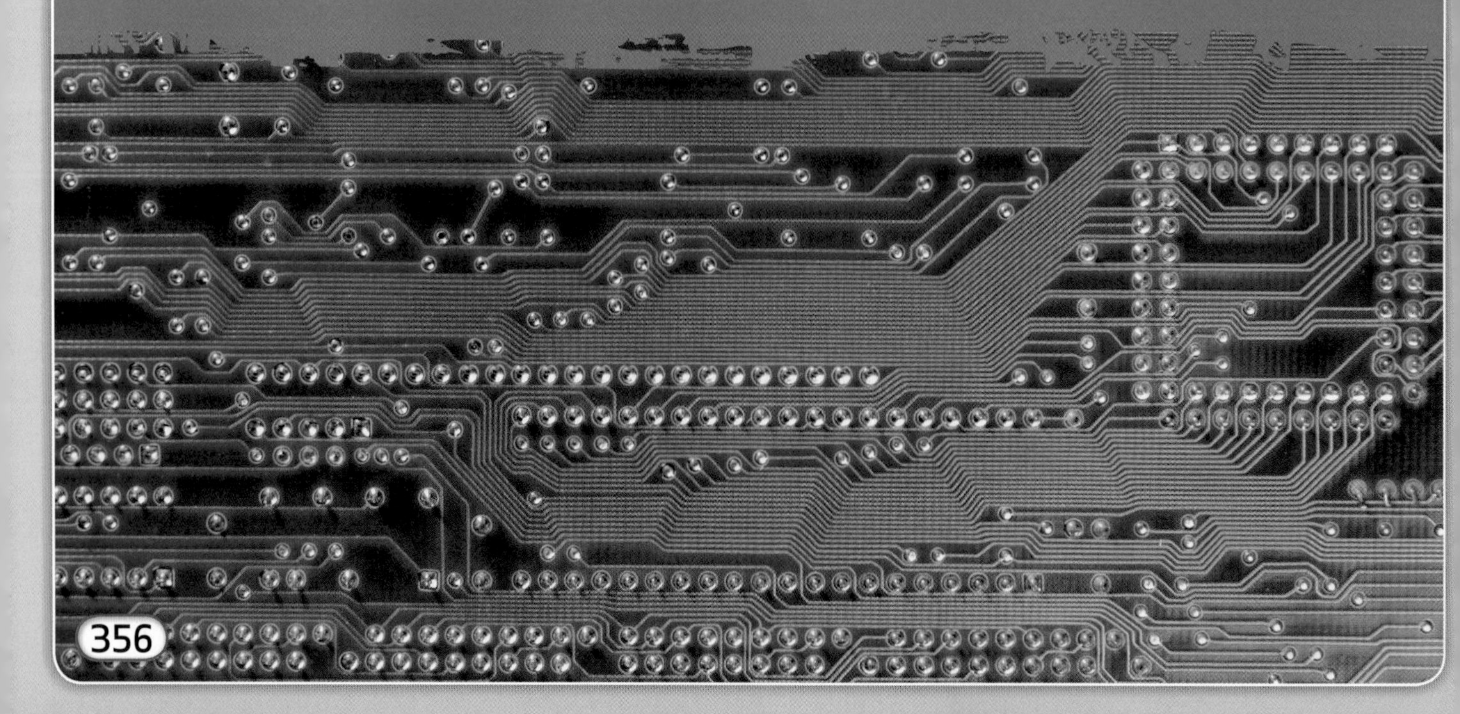

Glossary

A

Acceleration (ak-sel-er-Ā-shun)
Acceleration is a change in an object's velocity. (p. 261)

Artery (AR-ter-ē)
An artery is a blood vessel that carries blood away from the heart. (p. 64)

Atom (A-tum)
An atom is the smallest piece of matter that can still be identified as that matter. (p. 210)

B

Behavior (bē-HĀV-yor)
Behavior is any way that an animal interacts with its environment. (p. 16)

C

Chemical energy (KEM-i-kul EN-er-jē)
Chemical energy is energy that is stored in substances. (p. 298)

Circuit (SIR-cut)
A circuit is a looped path of conductors through which electric current flows. (p. 328)

Climate (CLĪ-mit)
Climate is the pattern of weather over a long period of time. (p. 170)

Communication (com-MYŪ-ni-CĀ-shun)
Communication is any behavior that lets animals share information. (p. 22)

Conductor (kon-DUK-ter)
A conductor is a material through which electricity can flow easily. (p. 326)

Current electricity (KUR-ent ē-lek-TRIS-it-ē)
Current electricity is a form of electricity in which electric charges move from one place to another. (p. 322)

D

Dwarf planet (DWORF PLA-nit)
A dwarf planet is an object that orbits the sun, is larger than an asteroid and smaller than a planet, and has a nearly round shape. (p. 129)

E

Electricity (ē-lek-TRIS-it-ē)
Electricity is a form of energy that involves the movement of electric charges. (p. 322)

Evaporation (ē-va-por-Ā-shun)
Evaporation is a change from the liquid to the gaseous state. (p. 157)

F

Force (FORS)
A force is a push or a pull. (p. 250)

Copper is a good conductor of electricity.

Front (FRUNT)
A front is the boundary where two different air masses meet. (p. 168)

G

Galaxy (GA-luk-sē)
A galaxy is a star system that contains large groups of stars. (p. 112)

Gravity (GRA-vi-tē)
Earth's gravity is a force that pulls things to the center of Earth. (p. 252)

H

Habit (HA-bit)
A habit is a behavior that is learned through practice. (p. 19)

Humidity (hyū-MID-it-ē)
Humidity is the amount of water vapor in the air. (p. 153)

I

Instinct (IN-stinkt)
An instinct is an inherited behavior that an animal can do without ever learning how to do it. (p. 16)

Wasps build their nest by instinct.

Insulator (IN-sū-lā-ter)
An insulator is a material that slows or stops the flow of electricity. (p. 326)

K

Kidney (KID-nē)
A kidney is an organ that removes wastes and extra water from the blood. (p. 72)

Kinetic energy (ki-NET-ik EN-er-jē)
Kinetic energy is the energy of motion. (p. 284)

L

Learning (LERN-ing)
Leaning is a change in behavior that comes about through experience. (p. 17)

Liver (LIV-er)
The liver is an organ that produces bile to help digest fatty foods. (p. 70)

M

Mass (MAS)
Mass is the amount of matter in an object. (p. 206)

Mechanical energy (mi-CAN-i-kul EN-er-jē)
The mechanical energy of an object is its potential energy plus its kinetic energy. (p. 284)

Mixture (MIKS-chur)
A mixture is two or more kinds of matter put together. (p. 212)

Moon (MŪN)
A moon is a large rocky object that orbits a planet. (p. 120)

Motion (MŌ-shun)
Motion is a change in position. (p. 250)

Organ (OR-gen)
An organ is a structure that carries out a specific job in the body. (p. 58)

The heart is an organ that does the job of pumping blood throughout the body.

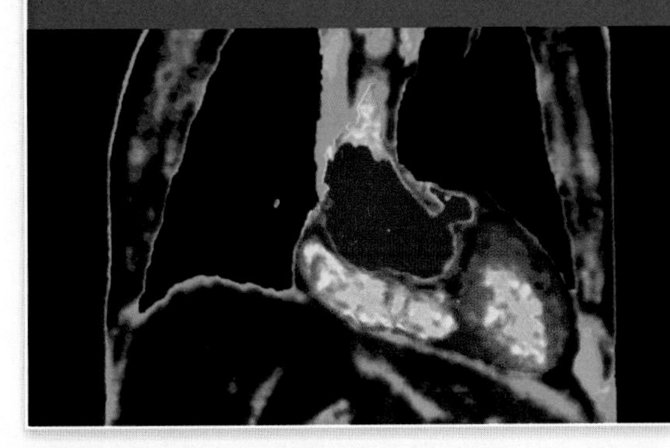

P

Pancreas (PĀN-krē-us)
The pancreas is an organ that helps digest proteins and starches. (p. 70)

Planet (PLA-nit)
A planet is a large nearly round space object that orbits a star. (p. 116)

Potential energy (pō-TEN-shul EN-er-jē)
Potential energy is stored energy. (p. 284)

S

Solution (so-LŪ-shun)
A solution is a mixture of two or more kinds of matter evenly spread out. (p. 216)

Star (STAR)
A star is a ball of hot gases that gives off light and other types of energy. (p. 110)

Static electricity (STA-tik ē-le-TRIS-it-ē)
Static electricity is a form of electricity in which electric charges collect on a surface. (p. 324)

Universe (YŪ-ni-vurs)
The universe is everything that exists throughout space. (p. 111)

Vein (VĀN)
A vein is a blood vessel that carries blood toward the heart. (p. 64)

Volume (VOL-yum)
Volume is the amount of space something takes up. (p. 208)

W

Water cycle (WAH-tur SĪ-cul)
The water cycle is the constant movement of Earth's water from the surface to the atmosphere and back again. (p. 156)

Weather (WE-thur)
Weather is the state of the atmosphere at a certain place and time. (p. 152)

The formation of clouds is part of the water cycle.

Index

T

Credits

Inside Front Cover (bg) NASA/Dimitri Gerondidakis. (fg) pbnj productions/Photodisc/Getty Images. (b) Joe Stancampiano/ National Geographic Image Collection. **ii–iii** (bg) Iakov Kalinin/Shutterstock. **iv–v** (bg) Jozsef Szentpeteri/National Geographic Image Collection. **v** (t) 81A Productions/ Photolibrary. **vi–vii** (bg) Digital Vision/Getty Images. **vii** (t) NASA, NOAO, ESA, the Hubble Helix Nebula Team, M. Meixner (STScI), and T.A. Rector (NRAO). **viii–ix** (bg) Sun Star/Getty Images. **ix** (t) Corbis/Jupiterimages. **x–xi** (bg) Robert Ranson/Shutterstock. **xi** (t) Victoria Pearson/Getty Images. **xii–1** Ludmila Yilmaz/Shutterstock. **2–3** (bg) Jupiterimages/Comstock Images/Getty Images. **3** (t) Cyril Ruoso/Minden Pictures/National Geographic Image Collection. (b) Jupiterimages/Getty Images. **4** (t) Renee Fadiman. (b) Tom Schwabel/National Geographic Image Collection. **5, 6–7** Stuart Westmorland/Image Bank/ Getty Images. **8** (t) Phil Degginger/Animals Animals. (c) ANT Photo Library/Photo Researchers, Inc.. (b) Michael Weber/imagebroker/Alamy Images. **9** (t) Mike McClure/ Index Stock/age fotostock. (b) clickit/Shutterstock. **10–11** (c) Jupiterimages/Comstock Images/Getty Images. **11** (tr) Bob Elsdale/Eureka/Photo Researchers, Inc.. (cr) Volker Steger/Photo Researchers, Inc.. (br) WILDLIFE/ Peter Arnold, Inc.. **12** (cr) Lightwave Photography Inc./Animals Animals. **12–13** (c) PATTY HERZOG/National Geographic Image Collection. **13** (tr) Stu_Naturally/Alamy Images. **14–15** (c) Marcos Veiga/Alamy Images. **15** (tl) Darylne A. Murawski/National Geographic Image Collection. (tr) Michael and Pataricia Fogden/Minden Pictures/National Geographic Image Collection. (bl) Paul Zahl/National Geographic Image Collection. (br) Christian Ziegler/National Geographic Image Collection. **16** (c) Phil Degginger/Animals Animals. **17** (t) Michael Weber/imagebroker/Alamy Images. **18** (t) Creatas/Jupiterimages. (b) Cathy Keifer/Shutterstock. **19** (b) Mike McClure/Index Stock/age fotostock. **20** (r) Cyril Ruoso/Minden Pictures/National Geographic Image Collection. **21** (t) Elzbieta Sekowska/iStockphoto. (c) Tom McHugh/Photo Researchers, Inc.. (b) ANT Photo Library/ Photo Researchers, Inc.. **22** (t) Michael Krabs/imagebroker/ Alamy Images. (b) Dave Watts/Alamy Images. **23** (b) clickit/ Shutterstock. **24** (b) Brain J. Skerry/National Geographic Image Collection. **25** (tr) Brian J. Skerry/National Geographic Image Collection. (br) Brian J. Skerry/National Geographic Image Collection. **26** (b) ©George McCallum/SeaPics.com. **27** (tr, b) Brian J. Skerry/National Geographic Image Collection. **28** (cl) Skip Higgins of Raskal Photography/Alamy Images. (b) Purestock/Alamy Images. **30** (b) Ed Reschke/Peter Arnold, Inc./Alamy Images. **32–33** (b) inga spence/Alamy Images. **33** (cl) James Forte/National Geographic Image Collection. (cr) J S Sira/Gap Photo/Visuals Unlimited. **34** (cl) Rick & Nora Bowers/Alamy Images. (bl) R B Forbes/ASM Mammal Image Library. **34–35** (b) Phil Degginger/Alamy Images. **35** (bl) FLPA/ - David Hosking/age fotostock. (br) Mark Moffett/Minden Pictures/National Geographic Image Collection. **36** (b) Purestock/Getty Images. **37** (c) Jason Lindsey/Alamy Images. (b) Matthew Polak/Corbis Sygma. **38** (cr) Sykes, P.W./U.S. Fish & Wildlife Service. (b) Lisa S. Engelbrecht / Danita Delimont,/Alamy Images. **39** (b) David Fleetham/Visuals Unlimited. **40** (tl) Jupiterimages/ Comstock Images/Getty Images. (tr) Dave Watts/Alamy Images. **40–41** (c) Digital Vision/Getty Images. **41** (b) Elzbieta Sekowska/iStockphoto. **42** (bl) Brant Allen. **43** (tl) Courtesy of Zeb Hogan (b) Sudeep Chandra/Zeb Hogan. **44** (tr) Tomasz Zachariasz/iStockphoto. (cl) Christian Ziegler/Minden Pictures/National Geographic Image Collection. **44–45** (b) Mark Moffett/ Minden Pictures/National Geographic Image Collection. **45** (tl) Tomasz Zachariasz/ iStockphoto. **46** (b) Mark Moffett/ Minden Pictures/National Geographic Image Collection. **47** (cl) Martin Dohrn/Nature Picture Library. (b) Mitsuaki Iwago/Minden Pictures/National Geographic Image Collection. **49** (cr) Andrew Darrington/ Alamy Images. **50** (b) Mark Moffett/Minden Pictures/National Geographic Image Collection. **51** (b) Dennis Kunkel Microscopy, Inc. / Phototake Inc./Alamy Images. **52** (b) Mitsuaki Iwago/ Minden Pictures/National Geographic Image Collection. **53, 54–55** (c) ejwhite/Shutterstock. **56** (t) Laurence Monneret/ Stone/Getty Images. (c) Andrew Northrup. (b) Symphonie/ Getty Images. **57** (t) Jupiterimages/Getty Images. (b) Corbis. **58–59** (c) Laurence Monneret/Stone/Getty Images. **59** (tr) GJLP/Photo Researchers, Inc.. **60–61** (c) David Epperson/ Photographer's Choice/Getty Images. **62–63** (c) Cultura/Alamy Images. **64–65** (c) Symphonie/Getty Images. **66–67** (c) Martin Strmiska/Alamy Images. **68–69** (c) 81A Productions/ Photolibrary. **70–71** (c) Jupiterimages/Getty Images. **72–73** (c) Corbis. **74–75** (c) Anders Rising/NordicPhotos/age fotostock. **77** (t) BananaStock/Jupiterimages. (ct) aleksandar zoric/Shutterstock. (c) Stockbyte/Getty Images. (c) Corbis. (cb) arenacreative/Shutterstock. (b) pjcross/Shutterstock. **78–79** (c) H. Mark Weidman/Workbook Stock/Getty Images. **79** (bl) George Grall/National Geographic Image Collection. (br) Jeff Hunter/Getty Images. **80** (t) Christopher Elwell/ Shutterstock. (tc) Bill Draker /Rolf Nussbaumer (tb) Michal Boubin/iStockphoto.Photography/Alamy Images. (bt) Mariya Bibikova/iStockphoto. (bc) Kurt_G/Shutterstock. (b) Andreas Gradin/Shutterstock. **81** (c) Runk/Schoenberger/Grant Heilman Photography/Alamy Images. **82–83** (c) Tony Heald/ Nature Picture Library. **83** (bl) Michael Durham/Minden Pictures/National Geographic Image Collection. (br) GeoTravel/ Alamy Images. **84** (bl) Vilmos Varga/Shutterstock. **84–85** (c) GFC Collection RF/Alamy Images. **85** (tr) Cathy Keifer/Shutterstock. (br) Kevin Dyer/iStockphoto. **86–87** (c) SMC Images/Getty Images. **87** (b) Petronilo G. Dangoy Jr./Shutterstock. **88** (tr) Jon Gilbert Fox. **89** (bl) Dr. Lori Alvord. (br) Stock Connection Distribution/Alamy Images. **90** (b) Reuters/Corbis. **90–91** (c) Quinn Rooney/Getty Images. **92** (b) Thomas Frey/imagebroker/Alamy Images. **93** (b) Robert Oliver/epa/Corbis. **94** (c) Mike Banks/Alamy Images. **96–97** (c) Marius Becker/epa/Corbis. **98** (tr) Martinez De Cripan/epa/Corbis. **98–99** (c) Hugh Gentry/Reuters/Landov. **99** (b) Thomas Frey/imagebroker/Alamy Images. **100** (b) Thomas Frey/imagebroker/Alamy Images. **101** Cotton Coulson/National Geographic Image Collection. **102–103** (bg) NASA, ESA, and The Hubble Heritage Team (STScI/AURA). **103** (b) Todd Gipstein/National Geographic Image Collection. (t) NASA Goddard Space Flight Center Image by Reto Stöckli (land surface, shallow water, clouds). Enhancements by Robert Simmon (ocean color, compositing, 3D globes, animation). Data and technical support: MODIS Land Group; MODIS Science Data Support Team; MODIS Atmosphere Group; MODIS Ocean Group Additional data: USGS EROS Data Center (topography); USGS Terrestrial Remote Sensing Flagstaff Field Center (Antarctica); Defense Meteorological Satellite Program (city lights). **104** (bg) Carsten Peter/National Geographic Image Collection. (inset) Jim Webb Photography.

105, 106–107 NASA, ESA, and E. Karkoschka (University of Arizona). **108** (t) T.A. Rector/University of Alaska Anchorage, T. Abbott and NOAO/AURA/NSF. (c) Jim Richardson/National Geographic Image Collection. (b) NASA, ESA, and The Hubble Heritage Team (STScI/AURA). **109** (t) NASA/JPL. (c) Stockbyte/Getty Images. (b) NASA, ESA, H. Weaver (JHU/APL), A. Stern (SwRI), and the HST Pluto Companion Search Team. **110–111** Jim Richardson/National Geographic Image Collection. **112** (inset) NASA, The Hubble Heritage Team and A. Riess (STScI). **112–113** (bg) NASA, ESA, A. Aloisi (STScI/ESA), and The Hubble Heritage (STScI/AURA)-ESA/Hubble Collaboration. **113** NASA, ESA, and The Hubble Heritage Team (STScI/AURA). **114** (l) John R. Foster/Photo Researchers, Inc.. (r) Detlev van Ravenswaay/Photo Researchers, Inc.. **115** Scott Johnson/Starfire Studios. **116–117** BSIP/Photo Researchers, Inc.. **118** USGS/Photo Researchers, Inc.. **119** NASA/JPL. **120** (t) Stockbyte/Getty Images. (b) NASA Goddard Space Flight Center Image by Reto Stöckli (land surface, shallow water, clouds). Enhancements by Robert Simmon (ocean color, compositing, 3D globes, animation). Data and technical support: MODIS Land Group; MODIS Science Data Support Team; MODIS Atmosphere Group; MODIS Ocean Group Additional data: USGS EROS Data Center (topography); USGS Terrestrial Remote Sensing Flagstaff Field Center (Antarctica); Defense Meteorological Satellite Program (city lights). **121** (bg) Jupiterimages/Brand X/Alamy. (inset) NASA/JPL-Caltech. **122** NASA. **123** (t, b) NASA/JPL/Space Science Institute. **124** Science Source/Photo Researchers, Inc.. **125** NASA/JPL. **126–127** (l) NASA/JPL-Caltech/T. Pyle (SSC). **127** (r) NASA/JPL/JHUAPL. **128** (t) NASA, ESA, H. Weaver (JHU/APL), A. Stern (SwRI), and the HST Pluto Companion Search Team. (b) Victor Habbick Visions/Photo Researchers, Inc.. **129** NASA, ESA, J. Parker (Southwest Research Institute), P. Thomas (Cornell University), L. McFadden (University of Maryland, College Park), and M. Mutchler and Z. Levay (STScI)/Space Telescope Science Institute. **130–131** (bg) Ira Meyer/National Geographic Image Collection. **131** (t) Shigemi Numazawa/Atlas Photo Bank/Photo Researchers, Inc.. (b) Thomas J. Abercrombie/National Geographic Image Collection. **132** (l) uliyan Velchev/Shutterstock. (c) NASA. (r) Ira Meyer/National Geographic Image Collection. **132–133** (bg) NASA, NOAO, ESA, the Hubble Helix Nebula Team, M. Meixner (STScI), and T.A. Rector (NRAO). **133** NASA/Johns Hopkins University Applied Physics Laboratory/Carnegie Institution of Washington. **134** P. Parviainen/Photo Researchers, Inc.. **135** (tl) (c)1997 The Field Museum, GN88132_8c. (tr) Corrin Green. (b) Mark Henle/The Arizona Republic. **136–137** (bg) Kerrick James/Alamy Images. **137** (inset) Stephen Alvarez/National Geographic Image Collection. **138** Chris Butler/Photo Researchers, Inc.. **139** Shigemi Numazawa/Atlas Photo Bank/Photo Researchers, Inc.. **140** (bl, br) William K. Hartmann. **141**(t) Randy Olson/National Geographic Image Collection. (bl, br) William K. Hartmann. **142** Dan Durda (FIAAA, B612 Foundation). **144** Randy Olson/National Geographic Image Collection. **145, 146–147** Todd Gipstein/National Geographic Image Collection. **148** (t) BrandX/Jupiterimages. (c) Owen Franken/Corbis. (b) Jerry and Marcy Monkman/EcoPhotography.com. **149** (t) Sylvana Rega/Shutterstock. (b) Skip Higgins of Raskal Photography/Alamy Images. **150–151** Digital Vision/Getty Images. **152–153** (bg) Jeremy Woodhouse/Blend Images/Alamy Images. **153** Owen Franken/Corbis. **154–155** BrandX/Jupiterimages. **156–157** Sylvana Rega/Shutterstock. **158–159** John Foxx Images/Imagestate. **160–161** Jerry and Marcy Monkman/EcoPhotography.com. **162–163** (bg) Al Braden/Alamy Images. **164** (t-b) Patrick J. Endres/Visuals Unlimited. parema/iStockphoto. Jason Cheever/Shutterstock. saied shahin kiya/Shutterstock. Palto/Shutterstock. Visuals Unlimited. **165** Jeff Kietzmann/NSF/Lightroom/Topham/The Image Works, Inc.. **166–167** Nick Caloyianis/National Geographic Image Collection. **168** National Remote Sensing Centre Ltd/Photo Researchers, Inc.. **171** (t) WorldFoto/Alamy Images. (c) Skip Higgins of Raskal Photography/Alamy Images. (b) Hanne & Jens Eriksen/Minden Pictures. **172–173** PhotoDisc/Getty Images. **174** Patsy Lynch/FEMA. **175** Cameron Davidson/Alamy Images. **176** (inset) Lynsey Addario/National Geographic Image Collection. **176–177** (bg) Lynsey Addario/National Geographic Image Collection. **177** (inset) Mike Goldwater/Alamy Images. **178** (l) John Foxx Images/Imagestate. (c) National Remote Sensing Centre Ltd/Photo Researchers, Inc.. (r) Jeremy Woodhouse/Blend Images/Alamy Images. **178–179** (bg) Jules Frazier/Photodisc/Getty Images. **180** (bg) Sepi Yalda. **180–181** (bg) Iakov Kalinin/Shutterstock. **181** Sepi Yalda. **182** (inset) Ilene MacDonald/Alamy Images. **182–183** (bg) Jim McKinley/Flickr/Getty Images. **183** (inset) Alan Diaz/AP Images. **184** (t) Charles W. Luzier/Reuters/Corbis. **184–185** (b) PhotoStockFile/Alamy Images. **185** (br) NASA/Goddard Space Flight Center/Scientific Visualization Studio. **186–187** (bg) Digital Vision/Getty Images. **187** (inset) Jim Sigmon/Dallas Morning News/NewsCom. **188** (t) Sean Donohue Photo/Shutterstock. (b) China Photos/Getty Images. **190–191** (b) David Paul Morris/Getty Images. **191** (tl) Bradley C Bower/AP Images. (tr) Sun-Star, Jack Bland/AP Images. **192** Digital Vision/Getty Images. **193** David Woods/Corbis. **194** (t) Khoroshunova Olga Underwater/Alamy Images. (b) moodboard/Corbis. **194–195** (bg) Wothe, K./Peter Arnold, Inc.. **195** (t) David Madison/Jupiterimages. (b) Rob Walls/Alamy Images. **196** (t, b) John Livzey. **197, 198–199** Khoroshunova Olga Underwater/Alamy Images. **200** (t) Hampton-Brown/National Geographic School Publishing. (c) Mark Thiessen/Hampton-Brown/National Geographic School Publishing. **201** (t) David Lewis/iStockphoto. (b) Polka Dot Images/Jupiterimages. **202–203** Sun Star/Getty Images. **204** (t) DigitalStock/Corbis. (bl, bc, br) Artville. **205** (bg) Oleg-F/Shutterstock. (fg) Maria Stenzel/National Geographic Image Collection. **206** (fg) Hampton-Brown/National Geographic School Publishing. **206–207** (bg) NASA Human Space Flight Gallery. **207** (fg) Stockbyte/Jupiterimages. **208** (bl, br) Hampton-Brown/National Geographic School Publishing. **209** (t) Mark Thiessen/Hampton-Brown/National Geographic School Publishing. (bl) Clayton Hansen/iStockphoto. (bc, br) travis manley/Shutterstock. **210** (fg) Patrick Lane/Somos Images/Corbis. **210–211** (bg) Michael Dykstra/iStockphoto. **212** David Lewis/iStockphoto. **213** Mark Thiessen/Hampton-Brown/National Geographic School Publishing. **214** (bg) John Burcham/National Geographic Image Collection. (fg) Kim Fennema/Visuals Unlimited. **215** (bg) iStockphoto. (fg) AMe Photo/Getty Images. **216** Polka Dot Images/Jupiterimages. **217** (tl) Susan Trigg/iStockphoto. (tr) vita khorzhevska/Shutterstock. (bl) Image Club. (br) Robyn Mackenzie/iStockphoto. **218–219** (bg) Louie Psihoyos/Science Faction/Corbis. **219** (fg) Hampton-Brown/National Geographic School Publishing. **220–221** (bg) Gunter Marx/Alamy Images. **222** (fg) sjeacle/Shutterstock. **222–223** (bg) John Foxx Images/Imagestate. **223** (fg) Liviu Toader/Shutterstock. **224–225** (bg) Wothe, K./Peter Arnold, Inc.. **225** (l) imagebroker/Alamy Images. (c) Kapu/Shutterstock. (r) Monika23/Shutterstock. **226** Ian Cartwright/Jupiterimages. **227** Bianca van den Bos/iStockphoto. **228** (l) Patrick Lane/Somos Images/Corbis. (c) Brand X Pictures/Jupiterimages. (r) Digital Vision/Getty Images. **228–229** (bg) NASA Human Space Flight Gallery. **229** (fg) Mark Thiessen/Hampton-Brown/National Geographic School Publishing. **230** (fg) Tehshik P. Yoon. **230–231** (bg) Jason Reed/Getty Images. **231** (t) Tehshik P. Yoon.

(b) Sebastian Wagner/Getty Images. **232–233** (bg) Gunter Marx/Alamy Images. **233** (fg) Lester Lefkowitz/Getty Images. **234** (bg) PhotoDisc/Getty Images. (t) Steffen Foerster Photography/Shutterstock. (b) John Foxx Images/Imagestate. **235** Denis Selivanov/Shutterstock. **236** (t) Edward Kinsman/Photo Researchers, Inc.. (c) C Squared Studios/Photodisc/Getty Images. (b) Dave Saunders/Stone/Getty Images. **237** Panoramic Images/Getty Images. **238** (tl) Ingram Publishing/Superstock. (tc) Stockbyte/Getty Images. (tr) Corbis. (bl) PaulPaladin/Shutterstock. (bc) Johanna Goodyear/Shutterstock. (br) Dmitry Melnikov/Shutterstock. **239** James Balog/Getty Images. **240** Steve Allen/Brand X Pictures/Jupiterimages. **241** (bg) Robert McGouey/Alamy Images. (fg) Mark Joseph/Getty Images. **242–243** Robert Garvey/Corbis. **244** Lester Lefkowitz/Getty Images. **245, 246–247** Aleruaro/age fotostock. **248** (t) LWA/Getty Images. (b) Westend61 GmbH/Alamy Images. **249** (t) Stuart Leslie/Alamy Images. (b) Dana Bartekoske/iStockphoto. **250–251** LWA/Getty Images. **252–253** (bg) Chris Stein/Getty Images. (fg) Stuart Leslie/Alamy Images. **254** (fg) Artville. **254–255** (bg) Jean-Leon Huens/National Geographic Image Collection. **256** (bg) Alex Telfer Photography Limited/Getty Images. (fg) The London Art Archive/Alamy Images. **257** (t) Steven Newton/Shutterstock. (b) Plate I, Illustrating Law II from Volume I of 'The Mathematical Principles of Natural Philosophy' by Sir Isaac Newton (1642–1727) engraved by John Lodge (fl. 1782) 1777 (engraving) (see 136315) by English School (18th century) Private Collection/ The Bridgeman Art Library Nationality/copyright status: English/out of copyright. **258** (l) PhotoDisc/Getty Images. **258–259** (r) Westend61 GmbH/Alamy Images. **260–261** (bg) Dana Bartekoske/iStockphoto. **261** (fg) Joris van Caspel/iStockphoto. **262–263** (bg) Gary Cook/Alamy Images. **263** (t) Radius Images/Alamy Images. **264–265** (bg) Lynn Saville/Getty Images. **265** (fg) Corbis/Jupiterimages. **266** William A. Sands. **267** (t) Echo/Getty Images. (c, b) William A. Sands. **268–269** (bg) Karim Jaafar/AFP/Getty Images. **269** (fg) Ryan McVay/Getty Images. **270** Dominique Douieb/Getty Images. **271** (bg) Thomas Barwick/Getty Images. (fg) Corbis. **272** moodboard/Corbis. **273** (bg, fg) GoGo Images Corporation/Alamy Images. **274** Juice Images/Alamy Images. **275** Greg Trott/Getty Images. **276** moodboard/Corbis. **277, 278–279** David Madison/Jupiterimages. **280** (t, b) Hampton-Brown/National Geographic School Publishing. **281** (t) Pixtal/age fotostock. (b) Corbis/Jupiterimages. **282** (fg) Ron Chapple/Corbis. **282–283** (bg) E. Braverman/Getty Images. **284** (t, b) Hampton-Brown/National Geographic School Publishing. **285** Pixtal/age fotostock. **286** iStockphoto. **288** Michael Durham/Minden Pictures. **289** Ryan McVay/Getty Images. **290** (inset) PhotoAlto/Sandro Di Carlo Darsa/Getty Images. **290–291** (bg) Steve Wignall/iStockphoto. **292** Chepurnova Oxana/Shutterstock. **293** (t) Ivanagott/Shutterstock. **294** (bg) Michael Haegele/Corbis. **294–295** (t) Basement Stock/Alamy Images.

295 (tr) Chuck Franklin/Alamy Images. (b) David Greenwood/Getty Images. **96** Victoria Pearson/Getty Images. **297** Kuzma/Shutterstock. **298–299** (bg) Corbis/Jupiterimages. **299** (fg) Roman Sigaev/Shutterstock. **300–301** (bg) Mario Hornik/iStockphoto. **301** (fg) AP Images. **302** (fg) javarman/Shutterstock. **302–303** (bg) Franck Goddio/Hilti Foundation, photo: Jérôme Delafosse. **303** (t) EarthSat NaturalVue/Franck Goddio/Hilti Foundation. (b) Franck Goddio/Hilti Foundation, photo: Christoph Gerigk. **304** (tl) Hampton-Brown/National Geographic School Publishing. (tc) Victoria Pearson/Getty Images. (tr) Corbis/Jupiterimages. (bl) Chepurnova Oxana/Shutterstock. (br) Michael Durham/Minden Pictures. **304–305** (bg) John Foxx/Getty Images. **305** (fg) Barry Yee/Getty Images. **306** John Livzey. **307** (t) Nancy R. Cohen/Getty Images. (bl) John Livzey. (bc, br) Thomas Culhane. **308–309** (bg) Sharon Day/Shutterstock. **309** (fg) DigitalStock/Corbis. **311** (t) Lowell Georgia/Corbis. (b) Joerg Boethling/Peter Arnold, Inc.. **312–313** (bg) iStockphoto. **313** (fg) Mark Boulton/Alamy Images. **315** (t) GlowImages/Alamy Images. (b) Images&Stories/Alamy Images. **316** DigitalStock/Corbis. **317, 318–319** Steve Terrill/Corbis. **320** (t) Garron Nicholls/Getty Images. (c) Steve Terrill/Corbis. (b) Lourens Smak/Alamy Images. **321** (t) Serdar Yagci/iStockphoto. (c) Yury Kosourov/Shutterstock. (b) Hampton-Brown/National Geographic School Publishing. **322–323** Garron Nicholls/Getty Images. **324** Hampton-Brown/National Geographic School Publishing. **325** Lourens Smak/Alamy Images. **326** (tl) Serdar Yagci/iStockphoto. (tr) Yury Kosourov/Shutterstock. (bl) Daniel Tang/iStockphoto. (br) ID1974/Shutterstock. **326–327** gh19/Shutterstock. **328–329** Hampton-Brown/National Geographic School Publishing. **330–331** Ismael Montero Verdu/Shutterstock. **331** (c) Matthew Chattle/Alamy Images. **332** Artville. **333** (t) George Doyle/Getty Images. (ct) Judith Collins/Alamy Images. (cb) Westend61 GmbH/Alamy Images. (b) Stephen Coburn/Shutterstock. **334** Stephen Webel/iStockphoto. **335** (bg) fStop/Alamy Images. (inset) Rob Walls/Alamy Images. **336** (l) Joe Tree/Alamy Images. (r) Dennis MacDonald/PhotoEdit. **337** Gordon Warlow/Shutterstock. **338–339** David R. Frazier Photolibrary, Inc./Alamy Images. **340** (t) Peter Dazeley/Getty Images. (bl, br) NOAA/DMSP. **340–341** (bg) Robert Giroux/Getty Images. **341** (r) Influx Productions/Getty Images. **342** (inset) Photodisc/Getty Images. **342–343** (bg) Reuters/Corbis. **343** (inset) J. David Ake/AP Images. **344** (l) Judith Collins/Alamy Images. (cl) Stephen Webel/iStockphoto. (cr) Rob Walls/Alamy Images. (r) Gordon Warlow/Shutterstock. **344–345** (bg) Robert Ranson/Shutterstock. **345** (inset) fStop/Alamy Images. **346** (l) Rafael Angel Irusta Machin/Alamy Images. (r) Tara Teich. **346–347** (bg) Kheng Ho Toh/Alamy Images. **347** (tl, tr) Rafael Angel Irusta Machin/Alamy Images. (b) Tara Teich. **348** (b) Gift of Hewlett-Packard Company/Computer History Museum. **348–349** (t) Robert Dant/Alamy Images. **349** (b) Mike Kemp/Getty Images. **350** (t) H. Armstrong Roberts/ClassicStock/Alamy Images. (b) Technology And Industry Concepts/Alamy Images. **351** PhotoDisc/Getty Images. **352** Elaine Thompson/AP Images. **353** (t) TWPhoto/Corbis. (b) AP Images. **354** (l) Eric Fowke/PhotoEdit. (r) True Colour Films/Getty Images. **355** Philippe Lissac/Godong/Corbis. **356** PhotoDisc/Getty Images. **EM1** Serdar Yagci/iStockphoto. **EM2** ANT Photo Library/Photo Researchers, Inc.. **EM3** John Foxx Images/Imagestate. **EM7** BrandX/Jupiterimages. **EM13** Artville. **EM16** Ed Reschke/Peter Arnold, Inc./Alamy Images. **Back Cover** (tl) Maria Fadiman. (tc) Robert Clark/National Geographic Image Collection. (tr) John Livzey. (bl) Beverly Joubert/National Geographic Image Collection. (bc) NOAA/Getty Images. (br) Pauline Beugnies. (bg) NASA/Dimitri Gerondidakis.